About the Author

Jason Bradford has been affiliated with Post Carbon Institute since 2004, first as a Fellow and then as a Board Member. He grew up in the Bay Area of California and graduated from University of California – Davis with a B.S. in biology before earning his doctorate from Washington University in St. Louis, where he also taught ecology for a few years. After graduate school he worked for the Center for Conservation and Sustainable Development at the Missouri Botanical Garden, was a Visiting Scholar at U.C. Davis, and during that period co-founded the Andes Biodiversity and Ecosystem Research Group (ABERG). He decided to shift from academia to learn more about and practice sustainable agriculture, in the process completing six months of training with Ecology Action (aka GrowBiointensive) in Willits, California, and then founded Brookside School Farm. While in Willits, Jason also instigated the creation of Willits Economic LocaLization (WELL) and was on the board of the Renewable Energy Development Institute (REDI). For four years he hosted *The Reality Report* radio show on KZYX in Mendocino County. In 2009 he moved to Corvallis, Oregon, as one of the founders of Farmland LP, a farmland management fund implementing organic and mixed crop and livestock systems, where he worked until early 2018. He sits on the Economic Development Advisory Board for Corvallis and Benton County, and serves as an advisor for the OregonFlora Project based at Oregon State University. He lives with his family outside of Corvallis on an organic farm.

About Post Carbon Institute

Post Carbon Institute's mission is to lead the transition to a more resilient, equitable, and sustainable world by providing individuals and communities with the resources needed to understand and respond to the interrelated economic, energy, ecological, and equity crises of the 21st century.

Acknowledgements

Nate Hagens gets credit for instigating this report as he wanted something to use in his Reality 101 class at the University of Minnesota. College students Miranda Edwards and Ayana Ito assisted with early research and helped me see what basic concepts needed to be emphasized or explained in detail. The staff of Post Carbon Institute professionalized this work, from the big picture and structural suggestions of Richard Heinberg, graphic design and narrative framing input by Asher Miller, to the detailed editing by Rob Dietz and Daniel Lerch. I appreciate the time and thoughts provided by the interviewees: Michael Bomford, Wes Jackson, Kathryn Draeger, Kenneth Mulder, Mike Eaton, and Kelley Eicher. Their unique voices add a welcome depth and variety to the material, and some of them I've known and been inspired by for many years. Shannon Cappellazzi and James Cassidy of Oregon State University were instrumental in getting the soil health study done and making sure I interpreted the results correctly.

Download this report and supplementary materials at postcarbon.org/future-is-rural.

The Future is Rural: Food System Adaptations to the Great Simplification
By Jason Bradford

post carbon institute

Post Carbon Institute • Corvallis, Oregon, USA • postcarbon.org • resilience.org

Contents

Figures

Tables

Introduction

Today's economic globalization is the most extreme case of complex social organization in history—and the energetic and material basis for this complexity is waning.[1] Not only are concentrated raw resources becoming rarer, but previous investments in infrastructure (for example, ports) are in the process of decay and facing accelerating threats from climate change and social disruptions.[2] The collapse of complex societies is a historically common occurrence,[3] but what we are facing now is at an unprecedented scale. Contrary to the forecasts of most demographers, urbanization will reverse course as globalization unwinds during the 21st century. The eventual decline in fossil hydrocarbon flows, and the inability of renewables to fully substitute, will create a deficiency of energy to power bloated urban agglomerations and require a shift of human populations back to the countryside.[4] In short, the future is rural.

Given the drastic changes that are unfolding, this report has four main aims:

- Understand how we got to a highly urbanized, globalized society and why a more rural, relocalized society is inevitable.
- Provide a framework (sustainability and resilience science) for how to think about our predicament and the changes that will need to occur.
- Review the most salient aspects of agronomy, soil science, and local food systems, including some of the schools of thought that are adapted to what's in store.
- Offer a strategy and tactics to foster the transformation to a local, sustainable, resilient food system.

This report reviews society's energy situation; explores the consequences for producing, transporting, storing, and consuming food; and provides essential information and potentially helpful advice to those working on reform and adaptation. It presents a difficult message. Our food system is at great risk from a problem most are not yet aware of, i.e., energy decline. Because the problem is energy, we can't rely on just-in-time innovative technology, brilliant experts, and faceless farmers in some distant lands to deal with it. Instead, we must face the prospect that many of us will need to be more responsible for food security. People in highly urbanized and globally integrated countries like the U.S. will need to reruralize and relocalize human settlement and subsistence patterns over the coming decades to adapt to both the end of cheaply available fossil fuels and climate change.

These trends will require people to change the way they go about their lives, and the way their communities go about business. There is no more business as usual. The point is not to give you some sort of simple list of "50 things you should do to save the planet" or "the top 10 ways to grow food locally." Instead, this report provides the broad context, key concepts, useful information, and ways of thinking that will help you and those around you understand and adapt to the coming changes.

To help digest the diverse material, the report is divided into five sections plus a set of concluding thoughts:

- Part One sets the broad context of how fossil hydrocarbons—coal, oil and natural gas—transformed civilization, how their overuse has us in a bind, and why renewable energy systems will fall short of most expectations.
- Part Two presents ways to think about how the world works from disciplines such as ecology, and highlights the difference between more prevalent, but outdated, mental models.
- Part Three reviews basic science on soils and agronomy, and introduces historical ways people have fed themselves.
- Part Four outlines some modern schools of thought on agrarian ways of living without fossil fuels.
- Part Five brings the knowledge contained in the report to bear on strategies and tactics to navigate the future. Although the report is written for a U.S. audience, much of the content is more widely applicable.

During the process of writing this report, thought leaders and practitioners were interviewed to capture their perspectives on some of the key questions that arise from considering the decline of fossil fuels, consequences for the food system, and how people can adapt. Excerpts from those interviews are given in the Appendix section "Other Voices," and several of their quotes are inserted throughout the main text.

Globalization has become a culture, and the prospect of losing this culture is unsettling. Much good has arisen from the integration and movements of people and materials that have occurred in the era of globalization. But we will soon be forced to face the consequences of unsustainable levels of consumption and severe disruption of the biosphere. For the relatively wealthy, these consequences have been hidden by tools of finance and resources flows to power centers, while people with fewer means have been trampled in the process of assimilation. In the U.S., our food system is culturally bankrupt, mirroring and contributing to crises of health and the environment. We can rebuild the food system in ways that reflect energy, soil, and climate realities, seeking opportunities to recover elements of past cultures that inhabited the Earth with grace. Something new will arise, and in the evolution of what comes next, many may find what is often lacking in life today—the excitement of a profound challenge, meaning beyond the self, a deep sense of purpose, and commitment to place.

The temple complex of Angkor Wat, built during the 12th century in present day Cambodia, was part of the one of the largest cities of the ancient world with perhaps 1,000,000 inhabitants at its peak.[5]

Part 1: Our Predicament

Volumes have been written about the related crises of civilization and planet Earth, such as biodiversity loss, soil depletion, climate change, and declining access to cheap energy. What follows is a short tour of our modern predicament, with a focus on food systems. Our economic and political system is locked into the impossible—perpetual growth on a finite planet. It is difficult to imagine the world in the coming decades as growth inevitably flounders and economies shrink, but to grasp the notion of why the future is rural, and why a full-fledged reinvention of social life that places food at the center again is in store, we need to review the history of city development, explain the coming decline of energy availability, and explore food system vulnerabilities to changes in energy and climate regimes.

The Rise and Decline of Cities

Prior to the Industrial Revolution, the population of the largest cities was on the order of 100,000 to 1,000,000.[6] With rivers, canals, clay pipes, and cobblestone roads as the conduits for moving people and goods, cities could grow only so big. Most people (typically more than 90%) lived in the countryside, where they could access food and dispose of their waste directly in the environment. The modest surplus from this countryside living supplied cities with needed resources.[7] As of 2018, the world has over 1,000 urban areas (which may include multiple city jurisdictions) with over 500,000 people at an average density of 11,000 per square mile (or 17 people per acre). Thirty-seven urban areas have over 10 million inhabitants.[8]

To understand why cities are the size they are, it's useful to consider models of urban systems, which are akin to biological models of metabolism.[9] Infrastructure serves as a city's giant circulatory system, with steel pipes, electric wires, concrete roads, railroad tracks, and canals moving fluids, solids, and energy into and out of the city. The circulatory system of a modern city is very active, and it requires high energy inputs to keep the city-dwellers fed and prevent the build-up of wastes.

Insect biology provides another insight into city size. A widely accepted explanation for why insects aren't larger is their lack of an active respiratory system that exchanges air with a circulatory system. In contrast to animals with lungs, insects rely on diffusion of air through their tracheas. Although diffusion limits the size of insects, it has the advantage of not requiring metabolic energy to function (that is, air passively moves into their bodies; it doesn't have to be forced by pumping action). To complete the analogy, prior to the fossil fuel era, cities could only be "insect-sized," whereas with fossil fuels they have grown "as large as dinosaurs."

Part of the story of industrialization is the unprecedented separation of people from their means of subsistence. Globally, and for the first time in human history, more people are now living an urban life than a rural one.[10] Mass urbanization has been made possible by the prodigious exploitation of fossil fuels. Cities have always been wholly reliant on the capacity of rural areas to produce basic goods, most importantly food. But due to the concentrated energy in oil, with its ability to power heavy equipment and transport goods over long distances, cities have been able to achieve the scale they do today by drawing support from a land base often several hundred times their own area.[11]

Because urban dwellers don't regularly interact with the landscapes that feed them, it is easy for them to be unaware of the true nature of current living arrangements, and to believe that the trends of the past couple hundred years will continue indefinitely

(Figure 1).[12] Cities are known as places of wealth creation, and money is the primary way people in cities access resources. Look around a city and you see that those with money have social status and material goods, while those without money may literally live under a bridge and go hungry.

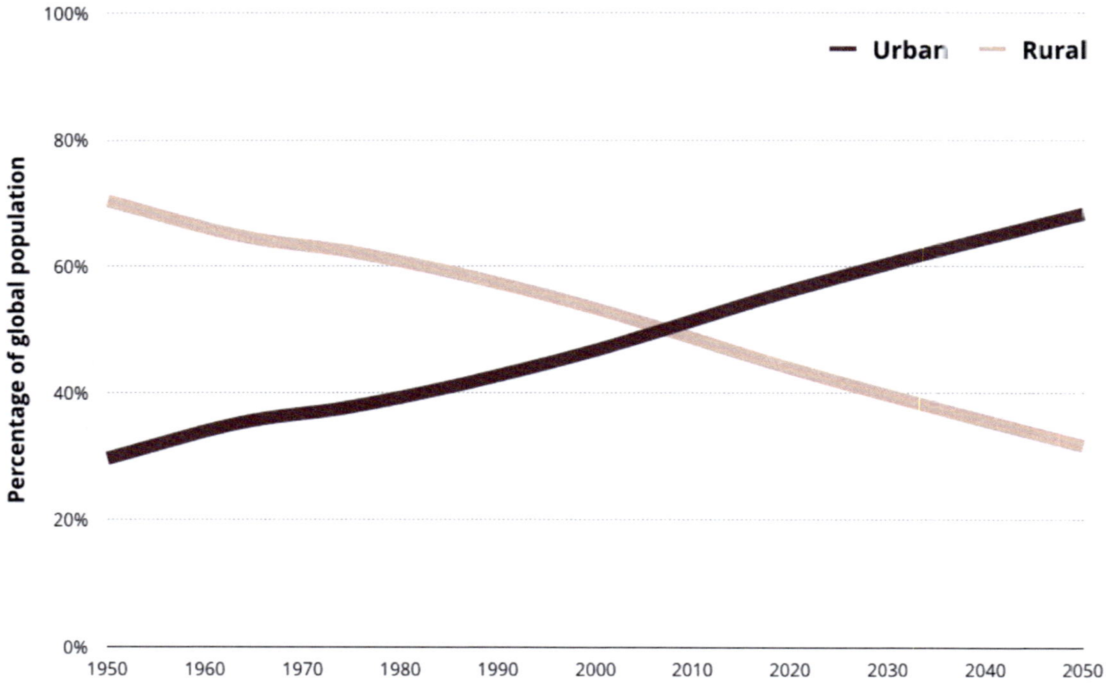

Figure 1. Percent of global population in rural vs. urban settings from 1950 to present, with United Nations projections to 2050.[13]

Real wealth consists of raw resources and finished goods and services. Although we've come to think of money as wealth, it's really just a socially accepted *claim* on real wealth. Because of the social construct of money, real wealth (those raw resources and finished goods and services) flows to people with access to finance. That's why finance may be considered a creative force, in that it can draw together the factors of production to overcome local deficiencies, which may lead to synergies that spur output. Cities, with their pools of finance, then, have become efficient places for bringing people together to share ideas, build trust, and innovate. Yet behind the curtain of finance, important flows of energy and materials are happening. Somewhere in the countryside, a barrel of oil is being extracted, a bushel of grain is being grown, and a cord of wood is being cut—physical things that no amount of money can create. As atmospheric scientist Timothy Garrett has noted, "Effectively, what sustains the purchasing power embodied in each one thousand dollar bill, and distinguishes it from a mere piece of paper, is a continuous 7.1 +/- 0.1 W {watts} of primary energy consumption."[14] In other words, money is worthless without the flow of energy that keeps industrial civilization humming.

As people have been removed from working landscapes and from nature, they have become deficient in the kinds of material experiences that normally defined the human condition. Lack of time outdoors interacting with the environment in physically constructive ways has led to gaps in mental models for how the world functions and distracted most of society from what we should be putting our minds to: namely, sorting out how to live without fossil fuels again.

The End of Cheap Energy

Whether because we can no longer afford the damage caused by burning them or because fossil hydrocarbons are a nonrenewable resource, the era of oil, coal, and natural gas will come to an end in the foreseeable future. It's the fossil fuel bonanza that has allowed modern cities, unlike in the past, to draw resources from areas well beyond their immediate surroundings by propping up global trade networks.

A key premise of this report is that renewable energy sources and technologies will not be able to compensate fully for the decline in fossil fuels. Society will have to get by on less energy and will have to use energy very differently, and these changes will have profound ripple effects on the economy, politics, and culture. I do not arrive at this premise casually, and the implications are certainly not easy to contemplate. The book and website *Our Renewable Future: Laying the Path for 100% Clean Energy* by Richard Heinberg and David Fridley provide a thorough explanation of the matter.[15] However, because energy literacy is so crucial for what is to follow, I will liberally reproduce a passage from the introduction to *Our Renewable Future*, "Why a Renewable World Will Be Different":

> Solar, wind, hydro, and geothermal generators produce electricity, and we already have an abundance of technologies that rely on electricity. So why should we need to change the ways we use energy? Presumably all that's necessary is to unplug coal power plants, plug in solar panels and wind turbines, and continue living as we do currently.
>
> This is a misleading way of imagining the energy transition for six important reasons.
>
> 1. **Intermittency.** As we will see in chapter 3, the on-demand way we use electricity now is unsuited to variable renewable supplies from solar and

wind. Power engineers designed our current electricity production, distribution, and consumption systems around controllable inputs (hydro, coal, natural gas, and nuclear), but solar and wind are inherently uncontrollable: we cannot force the sun to shine or the wind to blow to suit our desires. It may be possible, to a limited degree, to make intermittent solar or wind energy act like fossil fuels by storing some of the electricity generated for later use, building extra capacity, or redesigning electricity grids. But this costs both money and energy. To avoid enormous overall system costs for capacity redundancy, energy storage, and multiple long-distance grid interconnections, it will be necessary to find more and more ways to shift electricity demand from times of convenience to times of abundant supply, and to significantly reduce overall demand.

2. **The liquid fuels problem.** As we will see in chapter 4, electricity doesn't supply all our current energy usage and is unlikely to do so in a renewable future. Our single largest source of energy is oil, which still fuels nearly all transportation as well as many industrial processes. While there are renewable replacements for some oil products (e.g., biofuels), these are in most cases not direct substitutes (few automobiles, trucks, ships, or airplanes can burn a pure biofuel without costly engine retrofitting) and have other substantial drawbacks and limitations.* Only portions of our transport infrastructure lend themselves easily to electrification—another potential substitution strategy. Thus a renewable future is likely to be characterized by less mobility, and this has significant implications for the entire economy.

3. **Other uses of fossil fuels.** Society currently uses the energy from fossil fuels for other essential purposes as well, including the production of high temperatures for making steel and other metals, cement, rubber, ceramics, glass, and other manufactured goods. Fossil fuels also serve as feedstocks for materials (e.g., plastics, chemicals, and pharmaceuticals). As we will see in chapter 5, all of these pose substitution or adaptation quandaries.

4. **Area density of energy collection activities.** In the energy transition, we will move from sources with a small geographic footprint (e.g., a natural gas well) toward ones with much larger footprints (large wind and solar farms collecting diffuse or ambient sources of energy). As we do, there will be unavoidable costs, inefficiencies, and environmental impacts resulting from the increasing spatial extent of energy collection activities. While the environmental impacts of a wind farm are substantially less than those from drilling for, distributing, and burning natural gas, or from mining, transporting, and burning coal, capturing renewable energy at the scale required to offset all gas and coal energy would nevertheless entail

environmental impacts that are far from trivial. Minimizing these costs will entail planning and adaptation.

5. **Location.** Sunlight, wind, hydropower, and biomass are more readily available in some places than others. Long-distance transmission entails significant investment costs and energy losses. Moreover, transporting biomass energy resources (e.g., biofuels or wood) reduces the overall energy profitability of their use. This implies that, as the energy transition accelerates, energy production will shift from large, centralized processing and distribution centers (e.g., a 500,000 barrel per day refinery) to distributed and smaller-scale facilities (e.g., a local or regional biofuel factory within a defined collection zone or "shed"), since the same amount of "feedstock" cannot be concentrated in one place. It also implies that population centers may tend to reorganize themselves geographically around available energy sources.

6. **Energy quantity.** As we will see in chapter 6, quantities of energy available will also change during the transition. Since the mid-nineteenth century, annual global energy consumption has grown exponentially to over 500 exajoules. Even assuming a massive build-out of solar and wind capacity during the next 35 years, renewables will probably be unable to fully replace the quantity of energy currently provided by fossil fuels, let alone meet projected energy demand growth. This raises profound questions not only about how much energy will be available but also for widespread expectations and assumptions about global economic growth.

---

*Mikael Höök et al., "Hydrocarbon liquefaction: viability as a peak oil mitigation strategy," *Philosophical Transactions of the Royal Society A* 372, no. 2006 (2014): 20120319, doi:10.1098/rsta.2012.0319. David Murphy, Charles Hall, and Bobby Powers, "New perspectives on the energy return on (energy) investment (EROI) of corn ethanol," *Environment, development and sustainability* 13, no. 1 (2011): 179–202.

Food System Vulnerabilities

Heinberg and Fridley's six reasons why renewable energy systems will not seamlessly power society—intermittency, the liquid fuels problem, other uses of fossil fuels, area density of energy collection activities, location, and energy quantity—can easily be applied to the food system, which in places like the U.S. is just as industrialized and fossil-fuel dependent as any other economic sector. We will look at how energy is used in today's food system and discuss input substitution quandaries. Agriculture is also one of the most environmentally damaging activities on the planet, and ironically its externalities undermine future capacity to produce food, especially in the context of climate change and energy decline. Regional hyper-specialization has local communities reliant on global trade for basic food, a situation that has only recently developed and that can't be maintained indefinitely.

Inputs and Substitution Deficiencies

The industrial food system has been steadily replacing human and animal labor and local markets with mechanization and globally traded commodities. Just about every step in the industrial food system is so energy-demanding that, by now, someone eating a meal in the U.S. is ingesting only one kilocalorie of food for about every 10 kilocalories spent getting that food to their plate (Figure 2). How that excessive energy budget breaks down by activity is informative. The most recent compilation for the U.S.[16] shows the following:

- Farm activities and the embedded energy of inputs such as fertilizers account for about 14% of the total.
- Processing and packaging, which gives us such convenience and allows food to be shipped globally, is another 25%.
- The energy spent by warehouses, grocery stores, cafeterias, and restaurants is about 29%.
- The remainder, a whopping 28%, is used by households to go shopping, keep food in refrigerators, and cook.
- Transportation is only about 4% of the total, but consider that much of the energy used in the food system, such as processing and warehousing, allows for transportation efficiency.

Analyses of the U.S. food system show trends of greater consumption of highly processed foods and reliance on appliances rather than manual labor, both of which tend to increase energy consumption.[17]

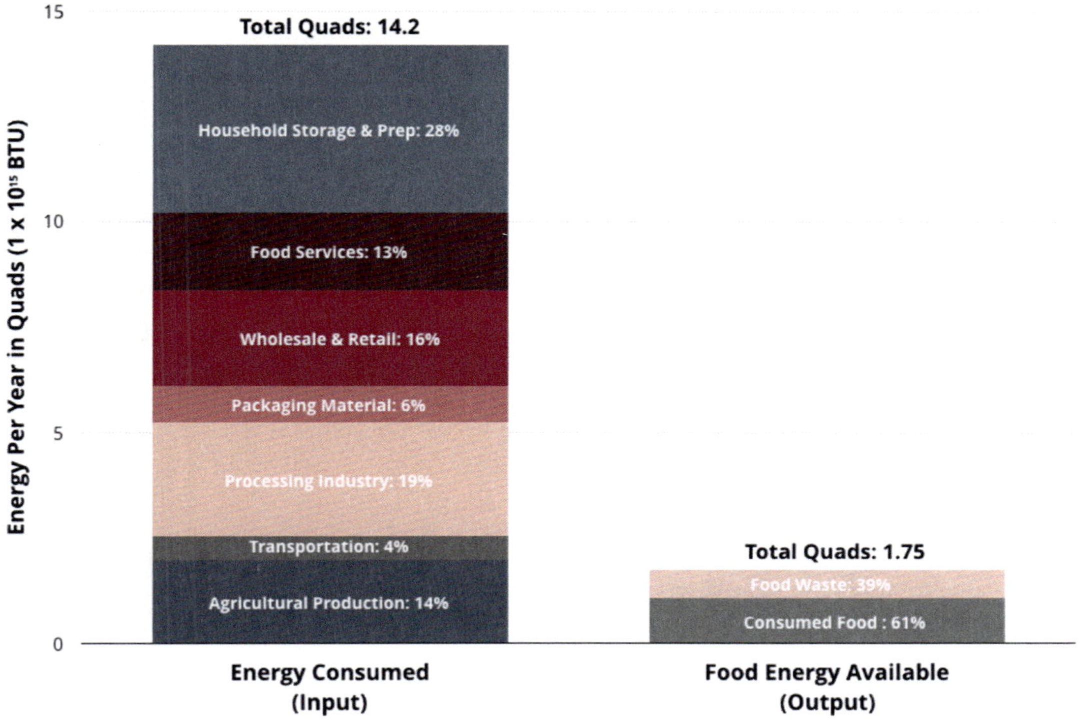

Figure 2. Energy input and output in the U.S. food system.[18]

This report can only provide a cursory overview of the complexities of petroleum, natural gas, and electricity dependency and substitution difficulty in the food system. For brevity sake, and to give a sense of the depth and scale of the challenge, I'll focus on one segment of the food system—agricultural production—and summarize how the six limiting factors from *Our Renewable Future* will affect this segment.

Let's begin with ***intermittency***. Farmers tend to wait for ideal weather conditions and then, as quickly as possible, prepare soil, plant seeds, and harvest crops. What they rely on today is liquid fuels (specifically, diesel and gasoline), delivered to on-farm storage tanks that fill large fuel tanks on tractors and harvesters. Farms can't afford to wait for the sun to shine or the wind to blow and hope that such an event corresponds to the right weather conditions for field activities. Although the intermittency problem could conceivably be solved by battery storage, this is not likely to work for many farm operations because of the low energy density of batteries.

Hence, we have ***the liquid fuels problem***: although you could theoretically run farms with electric-powered equipment, no technology known or likely to become available has the combination of transportability, storability, and high energy density that hydrocarbon liquid fuels offers. People often believe that because cars can successfully run on electricity, with battery packs allowing hundreds of miles between recharge, this same technology can apply to tractors. However, unlike cars running on smooth roads, typically at steady speeds, tractors are literally dragging steel through rough ground much of the time. Farm equipment tends to operate near its horsepower capacity, whereas a car might only work near capacity when accelerating into traffic now and then. Hydrocarbon liquid fuels are the only known substances with enough energy density that can be carried easily onboard a tractor under typical working conditions (e.g., a wide range of temperature; shaking and bouncing on rough terrain) and enable work to be performed continuously for many hours (Figure 3).

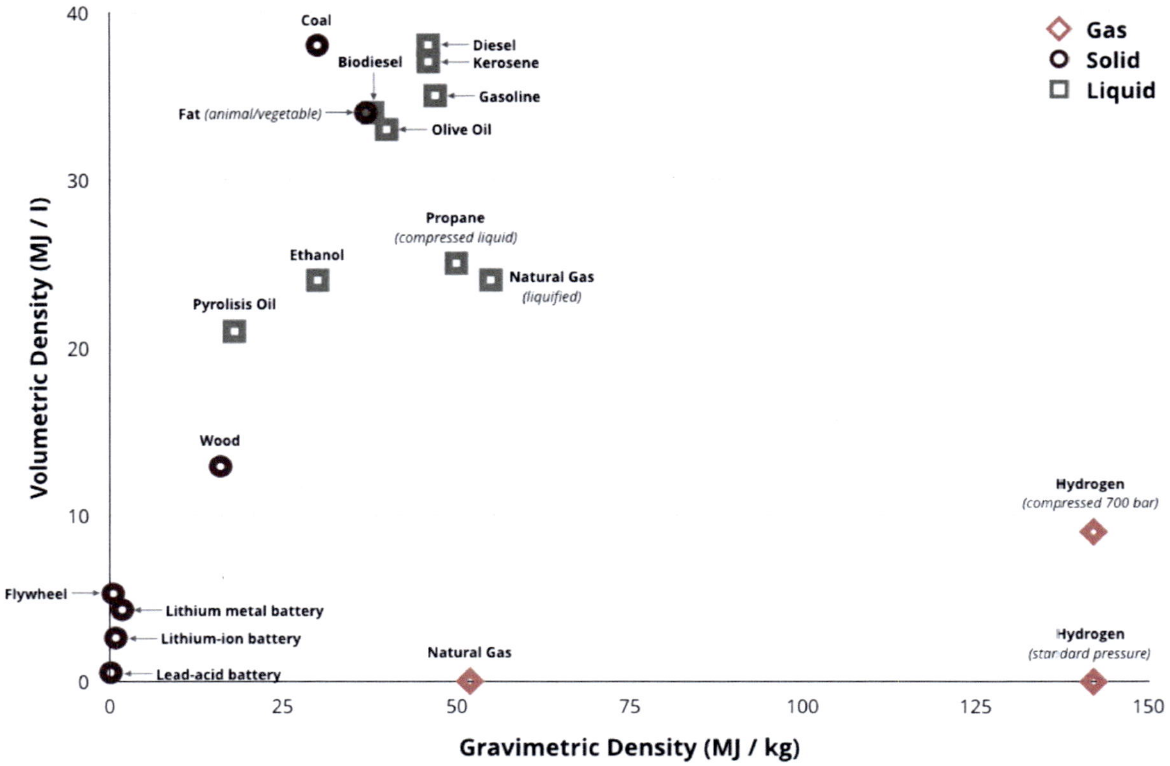

Figure 3. Energy density (gross heating value) of various storage forms is plotted by weight (MJ/kg) and volume (MJ/l).

An ideal energy storage source has both high gravimetric and volumetric density (upper right corner of graph). Alternatives to fossil fuels tend to be of lower density, making them more burdensome and costlier to use in general. Work performed will be less than gross heating value due to conversion inefficiencies, and efficiency can vary by the kind of work and how well the storage system is suited to it. For example, electric batteries to electric motors is a more efficient conversion (about 90%) than gasoline to internal combustion engines (about 30%). Batteries and other potential storage technologies can improve, but even at theoretical limits they would be many times less energy dense than fossil fuels.[19]

Other uses of fossil fuels include the feedstock for many products used on farms, such as pesticides and fertilizer. Plastics are becoming more abundant on farms as well, including irrigation pipes, weed suppression cloth, and roofing for greenhouses. While there may be ways to replace fossil-fuel-based supplies with renewable ones, such as crop-based feedstocks to make bioplastics and biofuels, these substitutes require using land that could otherwise be used to grow food.

Using land to yield renewable energy supplies also reveals the problem of ***area density of energy collection activities***. In some parts of the country it is common to see an oil well in the middle of a farm field. The oil well may occupy an area the size of a typical home to tap into a sizable underground reservoir. By contrast, if a farm needs to grow biofuel crops to power equipment, the area required to do so is going to be many times larger than the oil well.

Just as fossil fuel deposits are not evenly distributed around the Earth, renewable energy potential varies by ***location***. Furthermore, renewable energy sources are best used near their place of capture and storage. Farms tend to be located where soil and climate conditions are ideal. Some farms will be fortunately situated where great soils and rainfall patterns coincide with optimum solar radiation, consistent winds, or hydroelectric potential. But we have already reviewed why electricity will have a limited role in powering farms, even if it happens to be convenient to produce. The more certain renewable energy source on farms will be biomass. Farms of the future will need to make do with wood, straw, other crop residues, and extracted sugars or oils.

The ***energy quantity*** available to our society when powered by biomass plus renewable electricity will be far less that what we are accustomed to now. Coal and oil have higher energy densities than biomass, so they are more convenient and profitable to use.

I don't mean to imply that all troubles with agriculture and the food system began with the Industrial Revolution, and energy issues extend well beyond the farm. Even before tractors were invented, 19th century European and North American farmers became dependent on far-flung sources of guano to prop up depleted soils back home, with typical geopolitical maneuvers and tensions as a result.[20] With the advent of the Haber-Bosch process that uses natural gas to make nitrogen fertilizer, discussion of potential resource constraints today tends to concentrate on phosphate rock and how much is left to be mined in different parts of the world.[21] Both nitrogen and phosphorus supplies are reliant on abundant fossil fuels either as a feedstock or to power heavy mining equipment, and in transportation to farms, so ultimately the most crucial input to our current food system is the suite of products from the fossil hydrocarbon industry.

If oil supplies became unavailable suddenly on a large scale, most of the U.S. population would be at risk of going hungry within days or weeks, as just-in-time delivery systems and computerization have infiltrated the food system to a high degree.[22] Our complex and lengthy supply chains, which include getting fuel and fertilizer to farms, getting food to grocery stores, and even the free flow of money through bank accounts, are not things we should take for granted. The financial crisis of 2008 showed us how fragile these systems can be. While it is nearly impossible to plan for a global shutdown of commerce tomorrow, the fact that it is conceivable means we should be purposefully reducing our vulnerability as a matter of moral responsibility.

> *Where it makes sense, stop using so much machinery. Machinery is a really good thing when you want to produce more stuff. Machinery lets you do a lot more with less labor. But the reality is, we have enough stuff. Instead, let's start using our excess productivity to give people meaningful livelihoods. One thing I saw very clearly is that human power can be efficient, productive, and enjoyable.*
> *–Kenneth Mulder*

Externalities and Feedbacks

High-input food systems are not only vulnerable to supply shortfalls, they also generate pollution that undermines ecosystem productivity. Soil health can become impaired following years of tillage and recurring applications of synthetic fertilizer and pesticide. Fertilizers can degrade soil quality because they are made of acidic salts. And because fertilizers give plants freely available nutrients, the normal allocation of plant sugars to root exudates is reduced, severing the co-evolved relationship in which soil organisms make nutrients available to plants in exchange for energy.[23] Starved of its primary food source, the microbiome population declines, and the main path to building soil organic matter may be curtailed.

Pesticides, a catch-all term that includes herbicides, insecticides, fungicides, rodenticides, etc., can have non-target effects on the soil biome because, for example, an insecticide meant to kill an above-ground pest is likely to be harmful to soil insects too. Without a healthy soil biome, land will lack biological functions, such as efficient nutrient cycling, and becomes more dependent upon synthetic fertilizers to generate yields. The landscape surrounding farms suffers too, as agrochemicals get into water, drift in wind, and harm biodiversity beyond the field edge. Populations of birds, bats, wasps, spiders, and bees that interact with farms and provide cost-free services like pest regulation and pollination decline as chemical agriculture intensifies. The degraded landscape leads the farmer into more dependency on external inputs to compensate for undermined natural processes.

Farms export pollution via runoff into river systems. This pollution can have serious effects on fisheries and lead to direct human health hazards ranging from nitrates in drinking water to toxic algal blooms and oxygen-depleted oceanic dead zones (Figure 4). Greenhouse gases such as carbon dioxide, nitrous oxide, and methane are also excessive byproducts of industrial farming that exacerbate climate change. Ironically, an activity intended to contribute to human well-being and carrying capacity is undermining itself and the broader economy through multiple forms of pollution and unintended knock-on effects.

As the disciplines of environmental and ecological economics become more established, the externalized costs of industrial agriculture are becoming more recognized and quantified.[24] Adding up the damages makes industrial farming look economically poor for society even when it is profitable for the farmer. The environmental externalities of agriculture are so extreme that in many cases environmental costs are several times higher than the value of the resulting agricultural products.[25]

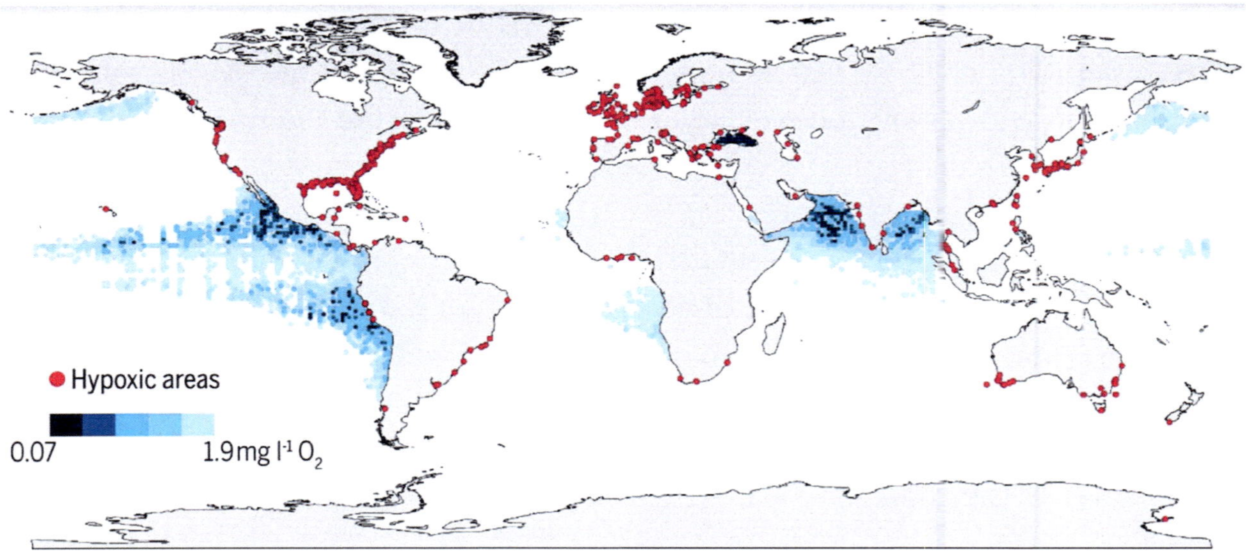

Figure 4. Ocean areas with severe oxygen depletion have increased dramatically since 1950, both in coastal and open ocean regions.[26]

How does a business survive when costs are higher than revenue? Mainly because many costs are being paid by someone else, and in many cases, are borne by future generations. The farm business sees current income and expenses, but what happens off the farm and in future years as damage isn't put on the books. For example, when a field erodes and leaches pollution into waterways, people downstream end up paying more to make the water drinkable, and those who depend upon high-quality water for their livelihood, such as fishermen, suffer. When a farmer uses herbicides and insecticides liberally over many years, native insect populations crash, and the food web erodes. Those who once relied on bees for pollination, or birds and bats for controlling pest outbreaks find themselves losing those services. When a farmer tills fields for annual, nitrogen-demanding crops year after year, the soil develops low rates of water infiltration and holding capacity, and therefore the landscape is unable to buffer rivers from extreme rainfall and snowmelt events. Rivers are more prone to flooding, and whole communities pay for the property damage and cleanup.

Plenty of experts have advocated changes to farming practices that would dramatically limit these liabilities, and these practices also happen to be ones that will be more resilient to erratic weather and fossil fuel depletion.[27] A crucial benefit of regenerative agriculture is that it can rebuild ecosystem service capacities, such as building soil, cleaning water, and pollinating crops, and thereby generate positive environmental outcomes.

I try to minimize, for example on my farm operation, the fertilizer usage. One of the pollutants of agricultural production is fertilizer nutrients leaching into the groundwater, so minimizing the appliance of the fertilizer prevents that. I'm literally farming the land that I live on: my home is on my farm, and underneath my farm is an aquifer that provides my drinking water and my animals' drinking water. I'm literally protecting my freshwater resource for the future.
– Kelley Eicher

Trade Dependency Risk

Export-driven commodity agriculture has led to regional crop specialization and a loss of local crop diversity. Whereas a farming community several decades ago in the U.S. may have produced crops and livestock mostly for local consumption and only a few for global markets, today the reverse is the norm. This means locally sourcing a balanced diet throughout the year is impossible for most of the population. Committed locavores eating their CSA share, raising backyard hens, and baking from locally milled flour are now the exception rather than the rule. Nearly everyone in highly industrial societies is in a similar position, making us vulnerable to any hiccup in global trade. History is replete with examples of trade relationships gone awry. If trade is for luxury items, an international disruption is no big deal. But nobody can afford to lose access to food. Civil unrest and wars can arise from such situations—either insufficient access to food, the land and water to grow it, or fertilizer and fuel supplies.[28]

While a rural population engaged in local trade relationships is less likely to be severely impacted by broader system disasters—whether financial, political, or natural—recovery from local disasters is very much aided by outside assistance. Localizing food is not about isolationism; it's about seeking balance between regional, national, and global systems of trade in the context of energy decline.

Those of us who grew up in a world fed by an industrial food system find it hard to imagine alternatives. The very old among us who were children prior to World War II, immigrants from countries without excessive energy use, and rugged travelers and Peace Corps workers know better what is possible. Our current living arrangement is a break from the past that took decades to develop. Hopefully the unwinding of industrial overkill is a process that extends over decades as well, so that we have time to develop elegant solutions to problems that will arise. Later sections of this report explore what such a food system might look like, perhaps evolving to be less industrialized given energy and environmental constraints, and how to start moving in that direction.

Regional Adaptation in the United States

Over 80% of the U.S. population live in urban areas. If the future is rural then where are people going to go? The answer to this question is two-fold: 1. based largely on climate and soil quality, some regions are more able to adapt than others, and 2. within regions, large cities will find it difficult to maintain their infrastructure and economies. Two experts on ecology and energy, John Day and Charles Hall, provide a regional view of long-term prospects for inhabiting North America:

> Rich natural ecosystems offer a fall back for the support of society as fossil fuels are depleted. Although society lived on the goods and services provided by natural systems for thousands of years, nature alone cannot sustain the energy intensive economic system that exists now. But at least nature is a cushion. If we learn to husband and protect natural systems, they offer a sustained flow of goods and services that society can use, and indeed has always used. Ecosystem services are high in the eastern US and in a narrow band of the Northwest. They are especially high in coastal zones and river valleys. Ecosystem services decrease from the southern Great Plains to the Southwest where they are lowest, and climate change will further reduce ecosystem services in this region. The high levels of ecosystem services in coastal regions will be impacted by climate change due to sea-level rise, stronger storms, and in some cases decreased freshwater discharge. Natural ecosystems have been degraded over wide areas of the earth. An important goal for society is restoring these systems so that they can continue to supply ecosystem goods and services.[29]

And further on they conclude:

> In summary, the emerging megatrends of the twenty-first century will result in large challenges for sustainability in the U.S. We believe that the most difficult areas to maintain are likely to be the southwest—especially southern California—coastal regions of the Gulf and Atlantic, and large urban regions especially those in the northeast, southern Florida, and the southern half of California.[30]

In addition to the regional perspective on ecosystem services across the U.S., a key point is the difficulty of supporting any *large urban region* as energy becomes more expensive. As a rule, population will be forced to disperse across the landscape more evenly. Smaller towns and cities surrounded by areas of high biocapacity will fare better than large cities and may be choice destinations for migrants. A local economy focused on basic goods and services everyone needs, such as food, energy, and fresh water, will outlast one based on non-essentials, such as financial services and tourism. As an example, Day and Hall suggest that the depressed, rust-belt city of Flint, Michigan, which is adjacent to the productive soils of the upper Midwest, has a far brighter future than the booming city of Orlando, Florida, which will be hammered by climate change and depends upon vacationers with disposable income. Those living in an area with poor prospects may consider relocating.[31]

The Great Simplification

In anthropological terms, as we have less energy available, our society will become *less complex*, characterized by fewer monetary transactions and an increase in subsistence and informal economies. The cultural implications are profound.[32] Progressively less energy from fossil fuels will require greater labor inputs and less reliance on mechanization over time. For a culture that mythologizes as progress the dominant trends of the 20th century, such as urbanization, financialization, and the replacement of labor with capital and machinery, this realization will come as a shock. The process outlined here will collectively be referred to as **the Great Simplification**, and corresponds to what permaculturalist and futurist David Holmgren calls the Energy Descent scenario:

> Energy Descent is the erratic but ongoing decline in the material and energy base to support humanity. In this scenario, as fossil fuels are depleted and the impacts of their past use continue (such as climate change), the nature of society will change to reflect many of the basic design principles if not details of pre-industrial societies. This will require a relocalisation of the economy, a re-ruralisation of settlements and reduction in the population that can be sustained in many countries. Novel technologies and cultural patterns may ease the transition but will not prevent the process of energy descent to less complex but more resilient ways to provide for human needs and values. As happened with many past civilisations (including the well documented decline of the Roman Empire), energy descent could occur through a series of precipitous crises that punctuate longer periods of stability.[33]

Food, its scarcity, the desire and opportunity to grow it, and the need to do it in ways that are appropriate to place and circumstance, will drive demographic shifts this century. People with life experiences and training aimed at urbanism are going to need a rapid education on what it takes to live off the land, and so-called conventional farmers and ranchers will have a steep learning curve to adopt more frugal and sustainable methods. But farmers and ranchers are not the only ones who need help adjusting to 21st century pressures. A society that aims to support them is also crucial to their success. If you are an educator, policy maker, someone who works in the food sector, or you have expertise in finance, energy, or transportation services, this report can help you think about how to nudge people toward a more thrifty and rural life, build and protect soils, and feed population centers with less energy.

In Focus: Reruralization in the U.S., Cuba, and Greece

Archeologists have recorded the rise and fall of cities, but most people don't think about a similar fate in modern times. And yet there are many instances of urban areas in population decline today, and the drivers and responses are illustrative of what can be expected more broadly during the Great Simplification.

For example, the so-called Rust Belt in the U.S. suffered the collapse of industrial activity. This removed the money supply needed to maintain populations and associated infrastructure. Urban agriculture has been one adaptive response by those who remain, a kind of reruralization in place that is culturally celebrated.[34]

For nations plugged into the global economy, sufficient financial capital allows them to import key resources, such as food and fuels. Reruralization has happened in countries as their ability to import basic goods was curtailed.

Cuba was linked to the global economy through its export of cane sugar within the Soviet bloc, and the country relied on Russian petroleum to run its farms. As the Soviet Union disintegrated politically, Cuba's entry into regional trade relationships was stymied by the U.S. With key energy and food imports dropping, Cubans quickly adopted programs to feed themselves during the so-called "Special Period," and many of these reforms persist today.[35]

Lacking sufficient fuel for tractors, farmers had to learn how to breed and work with oxen. Scarcity of synthetic fertilizer forced the adoption of organic agriculture techniques, such as using cover crops and making compost. Large, state-run, industrial farms were broken up and replaced by many small farms with decentralized management and agroecological programs. In towns and cities people established kitchen gardens and raised animals on food waste. Their society was held together by fostering a sense of shared sacrifice, including the implementation of food rationing. A high literacy rate and established networks to communicate face-to-face down to the neighborhood level may have also helped achieve a swift, effective response.

In another example across the Atlantic Ocean, the Greek government took on massive debts as it integrated into the E.U. economy. Following the 2008 financial crisis, these debts became untenable and forced a period of austerity that led to an economic depression. Many people left Athens and returned to small villages where they had family, citing very practical concerns: "When someone loses their job in a city and has no hope of finding another, they come here as a last resort. We will be the last to starve because when you have a field or a garden, you can produce food for yourself and make sure you survive."[36]

Harvesting and trucking green beans to a processing facility, which will package and eventually transport the beans to the frozen food section of grocery stores.

Part 2: Appropriate Ways of Thinking to Respond

Here's our predicament. Our entire way of life, even the food that sustains us, relies on a rapid drawdown of resources and is undermining environmental support systems. One of the steps we need to take is to clear out old assumptions and modes of thinking that developed within today's culture and adopt new mindsets that ground us, so we can adapt to novel circumstances.

This part of the report exposes some of the common beliefs that hinder proper responses, especially the myth of progress and its embedded techno-optimism. These beliefs offer a false hope that lulls many into passivity, unable to see things as they really are. Once blind spots have been exposed, it may become clear that there are no simple solutions, there are no brilliant experts solving these problems, and no significant political force is even aware of our predicament, let alone amassing the gumption to do much of anything useful right now.

We do have guiding lights, however. The science of ecology, and what is meant by "ecologically sustainable," teaches us about proper scale, that what is taken must be returned, and that nothing really goes away. We may act with more humility if we accept that nature is the containing system, and all parts of human society are subsystems. Resilience science and thinking are introduced as a core way to view the state of the world and guide its transformation. If something is unsustainable, it will not continue, and during those moments of breakdown come opportunities for change.

The Lull of the Progress Narrative

Post Carbon Institute Fellow Michael Bomford opens a chapter titled "Getting Fossil Fuels off the Plate" with a wonderful description of the relationships between plants, animals, and the history of life on Earth:

> I learned about photosynthesis in early grade school, but its implications didn't sink in for some time. When they finally did, I got excited.
>
> Suddenly I lived in a magical world filled with plants using energy from the sun to assemble themselves out of thin air. I was among the innumerable living beings interacting with one another on a solar-powered planet shaped by life itself. I could breathe because billions of years of photosynthesis had enriched my planet's atmosphere with oxygen stripped from carbon dioxide molecules. The carbon from those molecules had been reassembled into energy-rich chains that made up the bulk of living things and could be rendered to fuel my body. With every breath I took, my body released a little energy that had once been stored by a plant, reuniting carbon with oxygen to make carbon dioxide. Eating and breathing were photosynthesis in reverse. Without plants, I could do neither.[37]

I likewise marvel that the bulk of what plants and animals are made of is assembled from the air. Carbon dioxide is taken in directly by plant leaves, while nitrogen goes from the air into bacteria that convert it to a form plants can ingest via their roots. Carbon, nitrogen, oxygen, and hydrogen are well mixed elements in the atmosphere, while the heavier elements in all of us come from soils or, if we eat a lot of oceanic organisms, the salts of the sea. And the source of energy driving the animation of minerals in lifeforms is sunlight. To someone steeped in biology, it's obvious that we are one of many creatures, related and interdependent, on an amazing planet spinning through space.

Perhaps because I grew up in late 20th century American suburbia rather than on a farm, I needed a formal education to appreciate all of this. I have also come to feel there is a difference between being intellectually aware of something and experiencing it. I have the sense that people in America are aware that "you are what you eat" without having the experience to know it deeply and then consider the implications.

If we are what we eat, then what do we make of the fact that fossil fuels underlie our food system? [38] Beyond the graphs of energy inputs and calories supplied, we are creatures of the fossil-fueled world, not just in body but in mind. Energy has been so abundant and transformative that we take it for granted and assume it will always be cheap and unlimited. What this leads to is partial acceptance of a situation, where a problem is acknowledged but the proposal for solving it reveals energy illiteracy and clearly originates from a way of thinking steeped in the myth of progress, born of the fossil-fuel-fired society we inhabit.[39]

Many of us live in a world that is technologically dazzling. Whereas just about anyone can understand how a simple tool such as a hoe works, the gadgets we have today are too complex for most users to truly comprehend. And yet they work and keep getting better! This has led to an assumption that technology will continually advance and be deployed to solve anything.[40] Try pressing someone about environmental problems and, in the end, you will usually encounter the belief that we "have the technology," and someone will figure it out.[41] We see this with proposals by technophiles to tweak the food system.[42] Need to localize food production? Build skyscrapers designed to grow vegetables.[43] Need to replace fossil fuels at scale? Build a cellulosic biofuel industry.[44] Need to replace natural gas as a source for synthetic nitrogen fertilizers? Use wind energy and hydrogen instead.[45] Need to apply nitrogen more efficiently on corn? Buy a precision farming package.[46] Want to eat meat but avoid the environmental footprint of doing so? Create fake meat grown in industrial vats.[47] Must get that excess carbon out of the air? Devise "negative emission technologies."[48]

Some of these proposals may be helpful to a degree, but more often, they miss the point and reveal troubling aspects of our culture. In most schemes, reduction of labor (which is usually the most financially costly factor of production) and process optimization are emphasized over actually reducing resource dependency (look up "farming with robots" for example). Efforts to make the current agricultural system more resource-efficient and integrate renewable energy technologies are certainly welcome. But if our goal is to reduce energy and resource use at a scale that matters, while protecting and regenerating the soil, the current food system is structurally unable to do so for three key reasons:

1. The spatial relationships between fields, processing centers, and consumers makes it energetically and logistically intractable to cycle wastes back to the land;
2. Financial return on investment is the metric being optimized, so the incentives are not aligned with the goals; and
3. Efficiency gains in one process don't help reduce overall resource consumption as the broader economy simply responds by growing more—an economic conundrum that has been widely known since the early stages of the Industrial Revolution and is called Jevons' Paradox or the rebound effect (although in an eventual post-growth context it may not be as relevant).[49]

In many circumstances, efficiency is the wrong goal. We usually don't design critical systems to be maximally efficient but instead add redundancies and buffers. Commercial aircraft fly with duplicative monitoring and control systems. Bridges are over-engineered and built with structural safety margins. A bit of extravagance in the food system, not wastefulness but eating at higher trophic levels now and then, is something to be glad for as a society indulging a bit on steak, wine, and cream pie is not on the verge of famine and could afford to tighten its belt if necessary. Also, a resilient food system would emphasize crop diversity and grow things in an area even if they often grow better somewhere else, knowing that crop diversity is important for long-term soil health and therefore food security. Stewarding our soil is the kind of goal that should not be secondary to narrow notions of economic efficiency.

Ecology and Sustainability

Energy literacy and ecological literacy are closely related as the study of ecology includes understanding the flow of energy and materials through living systems. Ecology, chemistry, physics, and Earth science provide the basic principles of ecological sustainability, which will be briefly reviewed here.

The global economic system is subject to a structural constraint that has been studied extensively and explained very clearly by many thousands of scientists—namely, the Earth is finite, and no subsystem can be larger than the Earth system.[50] The economy, which functions as a subset of the Earth system, can become only so big. The Earth system is effectively closed with respect to materials (aside from the occasional meteorite, we're stuck with what we've already got on this planet), open with respect to energy, and with an energy budget given by the sun (supplemented in minor amounts by the heat of the Earth's core).[51] Fossil fuels allowed us to temporarily blow past the normal energy budget, but with severe consequences such as overheating and other pollution. The "Spaceship Earth" analogy is frequently used to explain limits. There can only be so many astronauts in a ship because the amount of food and energy for heating, cooling, and recycling of air and waste products is finite. Just like astronauts need to pay attention to their sources of food and energy and the ability of their machines to process wastes, humans can take only so much energy and material from the biosphere and rely on ecosystems only so much to process our waste before problems ensue.

Those already familiar with the concepts of ecological footprint, overshoot, planetary boundaries, carrying capacity, net energy, energy quality, and the approximate timing of peak energy flows may get the most out of this report.[52] With such ecological and energy literacy, readers are expected to have a realistic appraisal of the ability (or lack thereof) of renewable energy systems to substitute for fossil fuels at the scale required to keep our highly consumptive, hyper-industrial economy humming. Also, an understanding of what money is and the role of finance in resource extraction and distribution is helpful background.[53] Although we have reviewed some of the key resource dependencies and pollution processing vulnerabilities of the food system, this report will not develop the broader case for ecological overshoot, which should be self-evident by now. What it will do is describe what's required for an agricultural system to be ecologically sustainable, which is actually very simple to summarize:

- Minimize external inputs;
- Tightly cycle nutrients and prevent their leakage;
- Protect and renew soil health.

Of course, an ecologically sustainable agricultural system will not persist if the broader society isn't on board. Coupled to proper stewardship of farmland, a sustainable society would be thrifty with resources, understand regional carrying capacity, and apply much of its ingenuity in support of ecosystem services.

Resilience Science and The Adaptive Cycle

We can think of sustainability as a goal and resilience science as a framework to help move us in the right direction. The economy and food systems are "complex," meaning they are comprised of many parts interacting at different scales of place and time. Changes in one part may have unpredictable ripple effects throughout the system. A resilient system often can adjust to subtle shifts gracefully and may even be able to bounce back from temporary crises. But a resilient system is not necessarily a sustainable one. Our patently unsustainable food system has been quite resilient for decades, but because it undermines planetary life-support systems and depends upon the drawdown of limited resources, its time is running out. This report advocates a goal to transform our food system to a new state that is both sustainable and resilient.[54] Crises that can be foreseen can be prepared for, and systems can be developed and trialed to soften landings and shorten the duration of chaotic transition periods. In resilience parlance, we seek to move the food system into a new "stable state" that functions to feed people reliably as energy and other resources become scarce.

Fossil fuels are finite, so logically society should be considering how to make do without them. A graceful transition of our economy away from fossil fuels would require planning and implementation over decades, prior to any crisis developing. [55] Unfortunately, we've squandered that opportunity. The threat of climate change and fossil fuels' contribution to it have garnered plenty of attention, but even this has not been impetus enough to take substantial steps beyond talks and declarations of intent. It appears that the risk to business profits and government tax receipts is too significant a political barrier to avoid future suffering or the threat of environmental collapse.[56]

Applying resilience thinking, we can view the current system as being nested in a domain, where relationships within the system reinforce each other and create stability (Figure 5).[57] Fossil fuel dependencies are "locked in" and while we may want to shift the system to a new domain that doesn't rely so much on fossil fuels, the current system—like a ball in a basin—keeps moving back toward equilibrium as not enough force has been applied to move it past a threshold. The current system has been resilient, but it is highly unlikely to remain so for long. Those of us engaged in the work of preparing for the Great Simplification have recognized that we are about to cross thresholds, such as progressively less affordable fossil fuels and more frequent weather extremes, whether we want to or not.

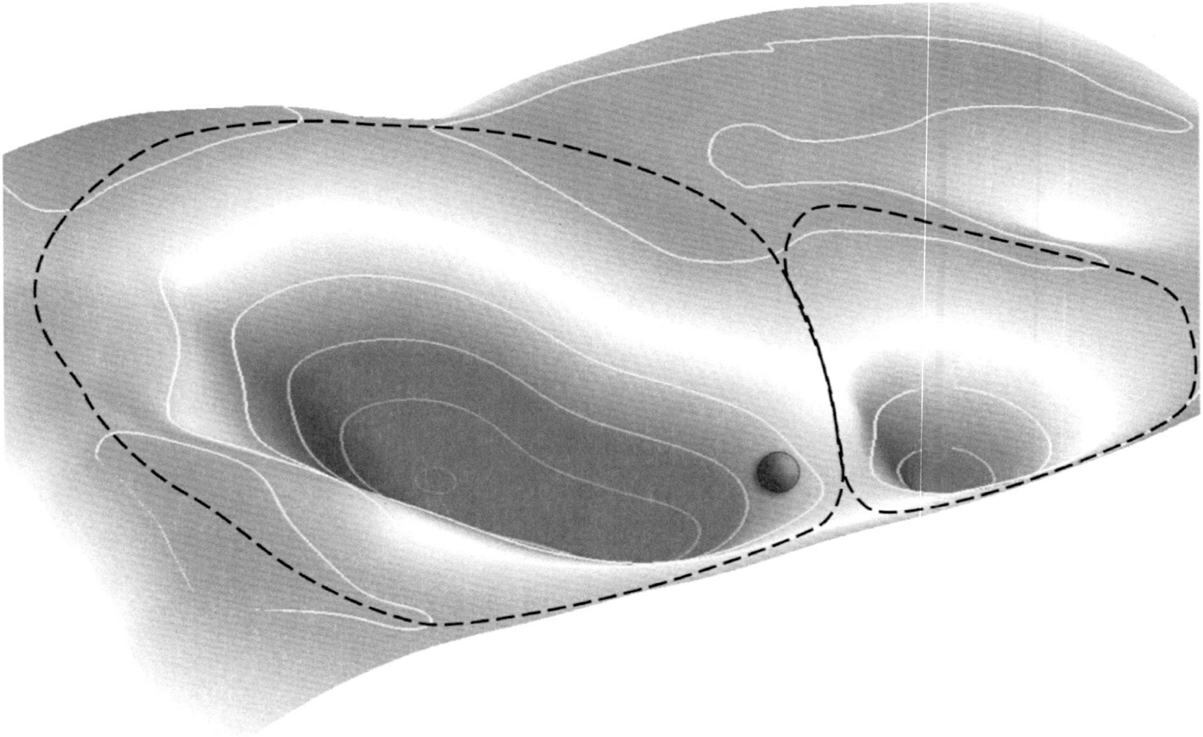

Figure 5. The system as a ball in a basin.
The ball is the state of the system. The basin in which it is moving is the system's domain; that is, it is the set of states that causes the ball to move toward equilibrium. The dotted line is a threshold separating alternate basins (or domains).[58]

A system under frequent stress and approaching thresholds may be at the late conservation phase of the fore loop of the adaptive cycle (Figure 6). Systems in this state tend to be robust, well connected, and resistant to change, but when they cross thresholds, the system may experience a great release (collapse). Such a release unleashes new opportunities through reorganization as the system settles into a new domain. We can use resilience planning to identify and promote actions today that will lead to desired outcomes during transformative, "back loop," periods. Before diving into what to do, let's review resilience science further to frame and guide our work.

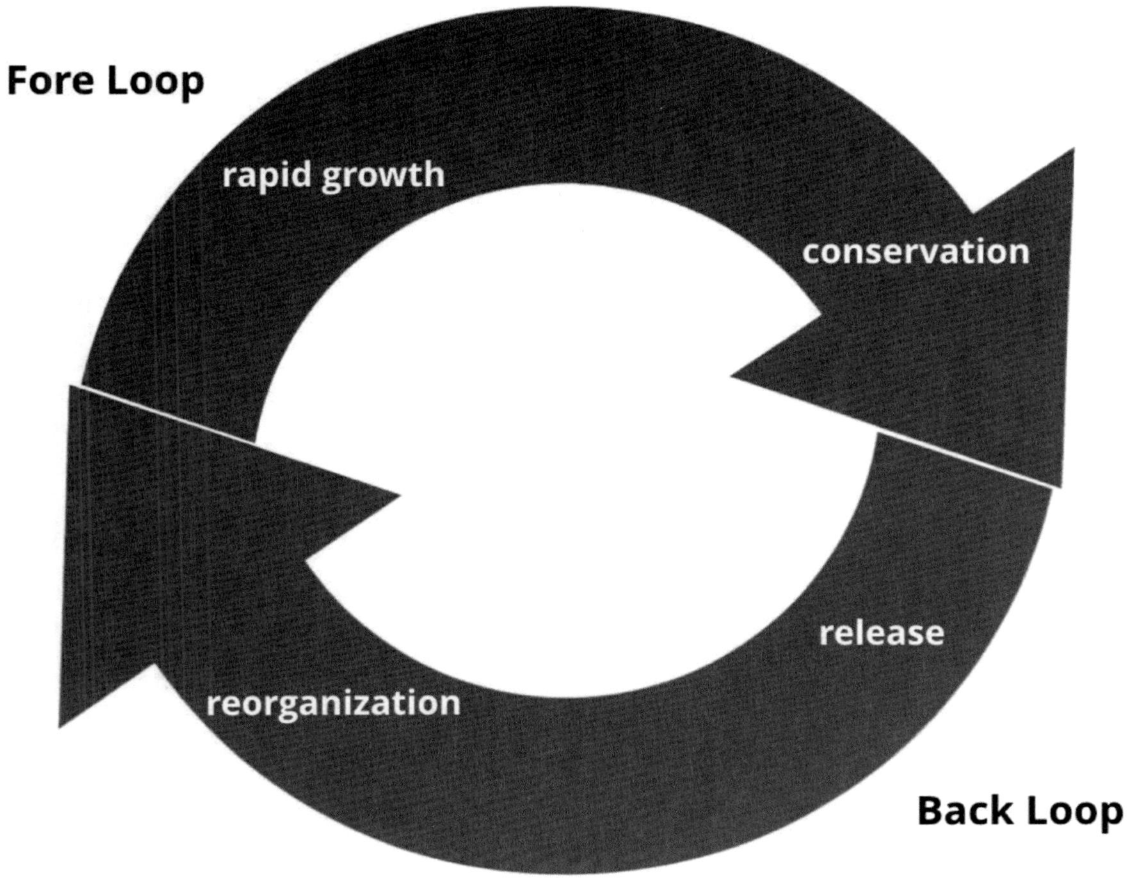

Figure 6. A simple representation of the adaptive cycle.
The rapid growth and conservation phases are referred to as the fore loop. This loop has relatively predictable dynamics and a slow accumulation of capital and potential through stability and conservation. The release and reorganization phases are referred to as the back loop. In this loop, which is characterized by uncertainty, novelty, and experimentation, there is a loss (leakage) of all forms of capital.[59]

Complex systems such as the global economy are comprised of many subsystems, or modules, such as the food system. Within the food system itself, there are nested sets of modules, such as individual farms, that are themselves following an adaptive cycle. For example, a multigenerational family farm that began as a homestead in the 1860s may have survived multiple transformations but is now shutting down and liquidating its assets because no remaining family members want to operate it. Perhaps the retiring farmer tried for many years to pass the farm on to other family members, but eventually gave up. This "giving up" was when the state of the farm the crossed the threshold from the domain of a viable farm with clear leadership to the domain of a farm that is dissolving. The "release" of an old farm allows other farming operations to grow as they acquire additional land and perhaps some of the equipment and people from the old farm. Modules are therefore parts of a complex system that interact with other parts—all these parts coexist within a set of hierarchical relationships that resilience scientists call a panarchy.

Panarchy is the conceptual framework that sees each module as having internal dynamics while also interacting with the broader system and influencing those dynamics. For example, an individual farm has relationships with fertilizer suppliers and customers for its products. The modules that sell this farm fertilizer or buy its crops are themselves buying and selling to others, such as natural gas feedstock for making urea or aggregating corn from area farms to run through a feed mill. If a crisis causes the loss of too many farms at once then fertilizer companies and feed mills may suffer and either adapt or dissolve themselves.

Soon the food system will go through a large-scale release phase. The framework of resilience science puts the challenge of this moment into perspective. Our job is to prepare to hasten reorganization and push the food system into a more sustainable domain. To prepare, we should be aware of potential stress events that test thresholds and think through possible leverage points in the system. In resilience terms we are promoting system *transformability*, i.e., recognizing that the current food system needs to change and creating awareness and options to help it do so.

New York is the most populous city in North America at about 8.7 million residents. When viewed as an urban region, New York is also part of one of the world's largest megacities with over 20 million inhabitants.[60]

Part 3: Key Knowledge for the Future of Food

If you and yours are going to have more responsibility for food in your community, then you will need to know how to get the most out of the resources in your area while maintaining their productive capacity over the long term. We can learn from how our ancestors operated (both in success and failure) before fossil fuels, and how those who live today without fossil fuels manage to do so. And we can also take advantage of the wealth of knowledge accumulated from modern scientific advances. The key areas you should have some basic knowledge about include:

- How local geography affects crop and livestock selection;
- Variation in adoption of industrial agriculture and urbanization;

- Shifting from a return on labor to a return on energy;
- Appreciating the multi-functional aspects of crops and livestock;
- Soil structure and processes;
- Nutrients and soil fertility management;
- Erosion control and soil conservation;
- The risks of intensification;
- Ecosystem services on farmland.

Geography and Crop/Livestock Selection

If we believe David Holmgren's premise that adaptations to the Great Simplification will likely reflect preindustrial human settlement patterns, then it's worth understanding how our ancestors evolved food systems as they spread around the planet and encountered diverse biomes. Ancient human cultures found ways to live within their local environments using only solar energy flows, so they knew some things that will be useful to us.

Agriculture characterized by extensive crop fields is a relatively recent invention. For most of human history people ate wild animals and plants. Horticulture developed as favored plants were brought into or adjacent to human camps and settlements. Mixed systems of gardening and hunting were common in tropical and relatively benign temperate climates. But not all places are suited to cultivating plants. In especially dry or cold environments, hunting and livestock husbandry dominated. Animals are especially important in higher latitudes for the storage of fat, ability to move with the seasons in nomadic cultures, and for byproducts such as leather and fur to keep people warm. Settled farming cultures revolve around domesticated seeds, mainly grasses, including wheat, rice, and maize. The grasses are typically paired with the seeds of legumes, such as lentils, pinto beans, and mung beans, which tend to have lower yields but are higher in protein than grains. Grains and legumes have some key properties that make them suitable foods for the rise of cities. Seeds typically have a high caloric density and dry on their own when ripe, which makes them rot-resistant in storage and easy to transport. For example, a pound of wheat contains about 1,500 kilocalories, whereas a pound of potatoes has only about 350. Getting high yields of grains requires fertile soil, which is why river valleys, where floods deposit deep, nutrient-rich soils, were the first locations of grain-based civilization.

Diverse small farms with limited mechanical support are complex operations that require a lot of intuitive systems thinking and a wide variety of physical activities. The work is hard but not arduous, and in my experience left time for leisure, art, music, community activism, and the mentoring of others. If one measure of a healthy life is the time spent not-sitting, mine was high on the chart! As a society we have this image of industrial farming as having liberated the bulk of us from a life of drudgery and boredom, but what it's done in fact is divorced us from nature and the cycle of seasons, and condemned most of us to sit.
–Mike Eaton

Certain animals were often kept close to dwellings to be fed leftovers. Pigs and chickens, for example, are monogastrics like us and have similar dietary needs (Figure 7). They are not picky eaters and can live off discarded kitchen scraps and the harvested crops that aren't quite good enough for making bread. The modern-day practice of growing grains for the "feed market" is a recent invention. Animals were domesticated to be our partners in managing our spoilage, to eat what we considered low-quality, and to consume the occasional bumper harvest to avoid a rat population explosion.

Even more dietary flexibility is exhibited by ruminants, such as cattle and sheep. Symbiotic populations of bacteria living in the guts of these livestock allow them to digest plant fiber and turn it into milk, meat, fat, bones, skin, and fur. Very little of the rural landscape is composed of things humans can readily eat, such as simple sugars, starches, fats, and protein, and most of what grows naturally is made of cellulose and lignin. Domesticated ruminants are how people found a way to turn the indigestible into the sweet and savory.

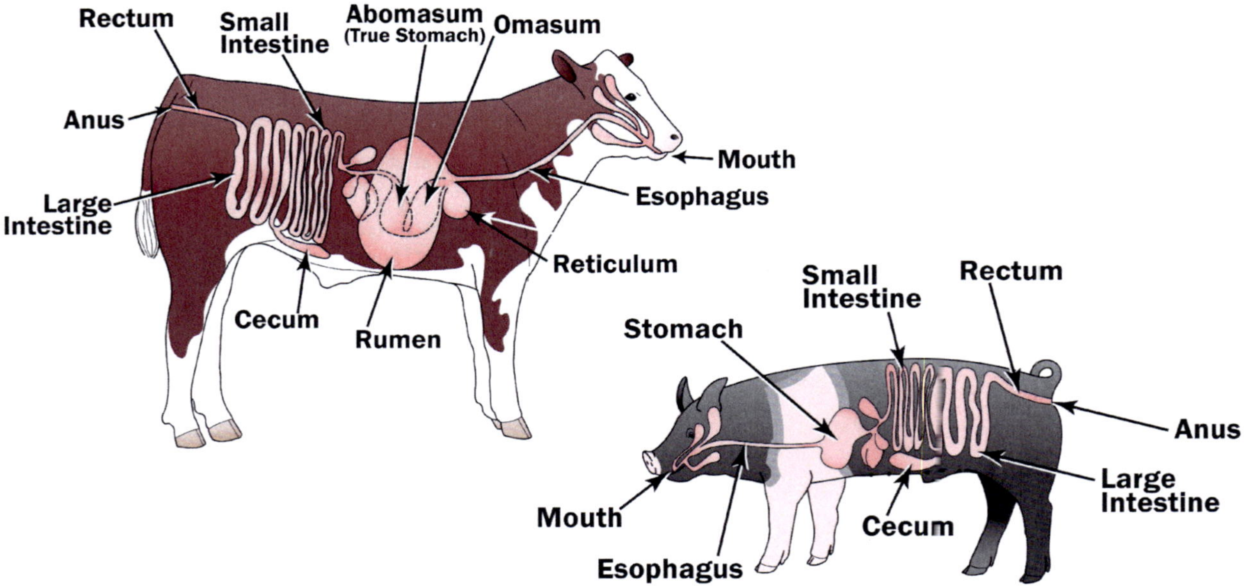

Figure 7. Digestive systems of ruminants (left) and monogastrics (right).[61]

Prior to the industrial revolution, work was performed in large part by people and their animals. Animals pulled tillage equipment and carts and helped with assorted tasks like grinding grain on mill stones or dragging lumber. Perhaps more important than this obvious kind of work, though, was the harnessing of livestock to improve field fertility.

Our primary crops are annual plants, meaning they are sown by seed and die within a year. To successfully grow these crops, at least the surface of the soil must be cleared of competing plants, traditionally through tillage, and sufficient soil nutrients must be available. Animals were often employed to shift and concentrate fertility. For example, livestock would graze extensively on hillsides or pasture fields during the day and then be moved into a field at night where they urinate and defecate. If night-time stocking is done for several weeks on the same field, it will become clear of competing vegetation, requiring little tillage prior to planting, and it will be chock full of nutrients from all the urine and manure. This is one way to grow a nutrient-demanding annual crop without mechanically applying synthetic fertilizer or hauling and spreading manure out of a barn. Farmers can manage the movement and concentration of fertility without livestock, but anyone who has built, turned, and spread a compost pile can appreciate the labor savings a flock of sheep can provide.

Successful human settlement of an area without significant trade requires not only producing sufficient food on average but being able to supply food security during lean seasons and years. Domesticated species of crops and livestock have been introduced around the world, and cultures have incorporated them to suit their needs. People have adapted their production techniques and achieved food security in a variety of places with gradients of land fertility and climate. A local food culture therefore evolves, and so do the genetics of human populations in response. Modern debates about what "should" be eaten usually fail to appreciate deep historical and environmental factors.

In the most benign tropical to subtropical environments, there is relatively little seasonality, and the growing season lasts almost all year. The need to store surplus food between seasons is low, so plants dominate the food supply, with perhaps a few chickens or pigs around to eat any food excess or waste. Certain tropical plants, such as palms, can provide dietary fat. A local vegan diet makes sense, given the type of agriculture in this climate.

In areas with plenty of tillable land, but more extreme seasonality, such as harsh winters or prolonged dry periods, storage of seasonal grain surplus is more critical. In addition to large granaries, more livestock are generally required than in tropical climates. To cope with harvest yield uncertainty, grains are over-sown, and any excess is shunted to livestock to act as a "feed buffer." If harvests are short, the livestock go hungry (and can be culled) before the people do. Animals convert grains into fats, which has 3,500 kilocalories per pound, more than double the caloric density of grains, and so is prized by people who perform physical labor. Dietary fat from livestock, which can store food without refrigeration, move on their own, and protect themselves from rodents and insects, is often more important than protein in these environments.

Most parts of Earth are not so-called breadbaskets with expansive, contiguous areas of deep topsoil and plentiful rainfall with which to grow crops. The landscape may only have pockets of good soil, which could be used to grow high-value tubers, vegetables, fruits, and some grains. But because of limited tillable area, livestock, especially dairy producers, may be an even more significant means of subsistence. And in the most extreme environments, including deserts and polar regions, nomadic herders, hunters, and foragers prevail.

Variable Food System Industrialization and Rural Populations

Today, the degree of adoption of industrial agricultural practices and participation in the global food system varies significantly around the world.[62] The most dedicated industrial systems are in countries with high per capita energy use and, perhaps consequently, relatively small rural populations, as fossil fuels are needed both to industrialize food systems and to support urban infrastructure (Figure 8).

Many people employ a mix of traditional and modern practices. For example, farmers who spread synthetic fertilizer by hand exhibit a minor turn toward industrialization. Substituting human and animal labor with more mechanization not only leads to decreases in rural population, but also comes with other consequences. Using more fuel and machines may increase labor productivity, but often comes with a decline in resource efficiency and an increase in pollution.

Prior to the Industrial Revolution nearly all countries would have been plotted in the upper left corner of Figure 8. By 2008, there were only 24 countries with a combined population of about 332 million people that were over 75% rural. In these countries, energy consumption averages less than one barrel of oil equivalent (BOE) per capita each year, which is about a fiftieth of U.S. levels. Over half the world's population, about 4 billion, lives in countries that are 50-75% rural, with average per capita BOE of 5.6. It first appears that only a modest addition of energy to an economy allows for substantial urbanization. However, in energetic terms, a barrel of oil represents around twenty thousand of hours of labor, or about a decade of work output by a human body.[63] So in countries with the seemingly modest consumption of five BOE per capita per year, on average everyone is leveraging the energy equivalent of 50 people working for them full-time.

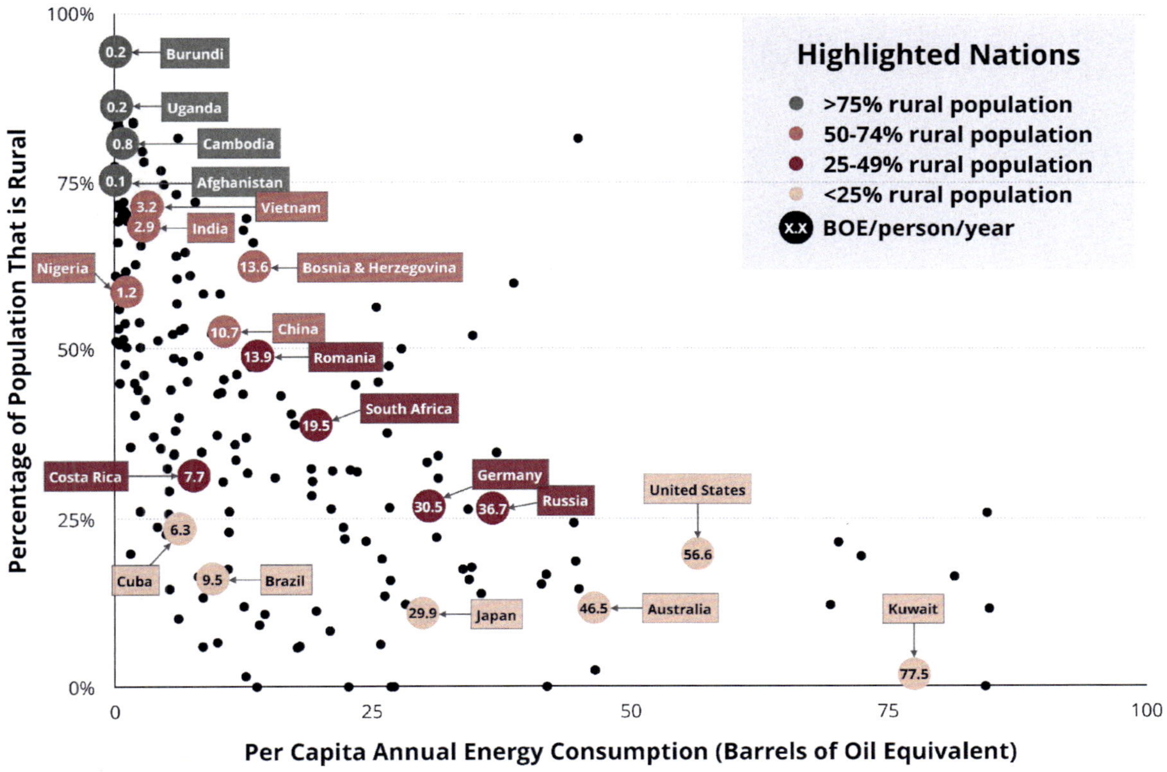

Figure 8. Per capita energy consumption and rural population by country (2008).
The percentage of population that is rural is plotted with respect to per capita energy consumption and shows that in general, countries with high energy use tend to be more urbanized. Some of the largest countries are highlighted, and outliers tend to be small island nations.[64]

The quartiles 25-50% and 0-25% of population living in rural areas use 18.6 and 32.2 BOE per capita per year respectively. The range of values by country is large, and likely reflects differences in latitude (tropical nations don't need to heat buildings), special circumstances (large petroleum exporters often have massive domestic consumption), variation in land use policies and transportation (e.g., the ability to live in the country but drive into the city for work), and in the era of globalization the ability to specialize in service industries such as finance and tourism and import most basic goods (e.g., Singapore).

It's questionable how substantially renewable energy systems will replace fossil fuels, but the evidence suggests renewables won't fully substitute and that total energy supply will decline. Unfortunately, nations aren't investing in renewables (or altering behavioral patterns) at a scale commensurate with the need to meet greenhouse gas reduction pledges.[65] But this does not necessarily spell disaster. Those in very high consumption countries can take solace in the fact that most of their consumption is probably wasted, and they could live on a fraction of what they do now, as most of the world still does. In the U.S., for example, GDP per capita in 1950 was about a third of today's.[66] Think of it this way: going on a restricted diet isn't easy but is usually a healthier way to live.

Return on Energy and the Return of Labor

The story we tell ourselves about modern farming and the food system—that it is incredibly efficient—may only be true through the lens of human labor. Seen through other perspectives, such as resource use and pollution, astonishing inefficiencies and atrocities abound. In 1972 the agricultural economist Michael Perelman wrote:[67]

> If we are facing an energy crisis, then we might do well to measure efficiency in terms of output per unit of energy instead of output per unit of labor, not only in agriculture but elsewhere in our economy.
>
> If we should decide to measure efficiency in terms of the conservation of energy, then American agriculture comes out very poorly. Harris estimated that Chinese wet rice agriculture could produce 53.5 BTU of energy for each BTU of human energy expended in farming it. For each unit of energy the wet rice farmer expends he gets more than 50 in return; for each unit of fossil fuel energy we expend we get about one-fifth in return. On the basis of these two ratios, Chinese wet rice agriculture is far more efficient than our own system.

During the Great Simplification, as the cost of energy goes up and economies stumble, the cost of labor will most likely go down. This scenario will make it more difficult to live in high-cost, urban settings, drive migration to rural areas, spur development of non-monetary exchange systems, and reestablish labor as a competitive and essential factor of production in farming. Beyond looking to the past, people in places like the U.S. may want to seek advice from those in countries that haven't fully industrialized. They demonstrate ingenuity that should delight us and just might keep us from going hungry.

Multi-Functional Crops and Livestock

One of the key lessons from a study of traditional agrarian systems is that crop and livestock choices are multifaceted. An unusual feature of today's food system, and an obvious consequence of extreme surplus, is that grains are grown specifically to feed livestock in amounts that exceed any historic norm. In peasant societies, livestock numbers were kept to what is termed "default" levels, where livestock were used for labor, to manage fertility, as clean-up artists, as feed buffers, and to harvest roughage.[68]

The preponderance of beef cattle, for example, is a modern invention. In the past, beef was a byproduct of old cows, male dairy calves, and retired oxen. Chicken meat was also the byproduct of old laying hens and young roosters. Historically, we see that animals served multiple functions to a greater extent than today.

Beef cattle and meat chickens are one example of a general rule. Specialization in agriculture has been taken to extremes in all kinds of crops and livestock. Dwarf wheat, for example, has very high seed yields, but is lacking in straw. If you are a specialized grain farmer and can buy fertilizer, the low straw yield may not seem important. For a diversified farmer using livestock and lacking cheap fertilizers, the old wheat variety may look better to you. That's because the abundant straw, when gathered, provides bedding for the barn and will turn into compost, and when left on the field, provides a welcome soil mulch that suppresses weeds and benefits the soil biology.

Interest in so-called heritage breeds of crops and livestock is not just nostalgic. Some people realize they have not had their multi-functional nature bred out of them and that such traits will be important again someday. Breeding programs for a wide array of crops and livestock that combine modern techniques with the needs of farming systems that function without fossil fuels would be very welcome in advance of the Great Simplification.

Soil Structure and Processes

In the U.S. today only a small percentage of the population is directly engaged in agriculture and ranching, and despite some good practitioners, overall management of soils and water is poor. History is riddled with examples of civilizations that progressively degraded soils to the point of food system collapse.[69] Perhaps we can forgive people for not understanding what they were doing, although plenty of documents suggest that many living at the time did. But we shouldn't give ourselves a pass for what is happening today as geology, archeology, soil science, and agronomy have progressed to the point where anyone can understand our predicament and the reasons for the sad state of most agricultural soils around the world. Only with the use of fossil fuels have we been able to temporarily postpone the reckoning by supplying the means to import lost nutrients and add synthetic nitrogen, and those won't last forever. Perhaps resource management would improve if people recognized that someday the ability of their families to eat will depend on the presence of healthy soils.

We've been burning highly dense carbon—the fossil fuels—to make our way of life possible. Well we've learned some important things, and we have scientific verifiability that we are stardust, and of the journey from minerals to cells on Earth and Darwinian selection, and so on. In thinking about that journey, we know that hunter-gatherers would not have been able to do what we've done. Early agrarian societies would not be able to do what we've done. They did not have the slack in their systems that fossil fuels provide.
–Wes Jackson

The creation of soil from rock and minerals is both a physical and biological process, called weathering, and soil itself is a mixture of inert minerals, pore space for air and water to move, living organisms, and dead parts of plants, animals, fungi, and bacteria in various states of decay. Soil minerals consist of particles that range in size—large in sand and very small in clay, with silt in between. Soils with high sand content don't store water and nutrients well but tend to have easy movement of air and water. Soils with high clay content are full of nutrients and can store a lot of water, but the clay holds onto water and nutrients tightly and doesn't allow for air movement, so plant roots don't grow well. Silt is more ideal for plant growth, and the very best farming soils, called loams, have a balance of silt, sand, and clay (Figure 9).

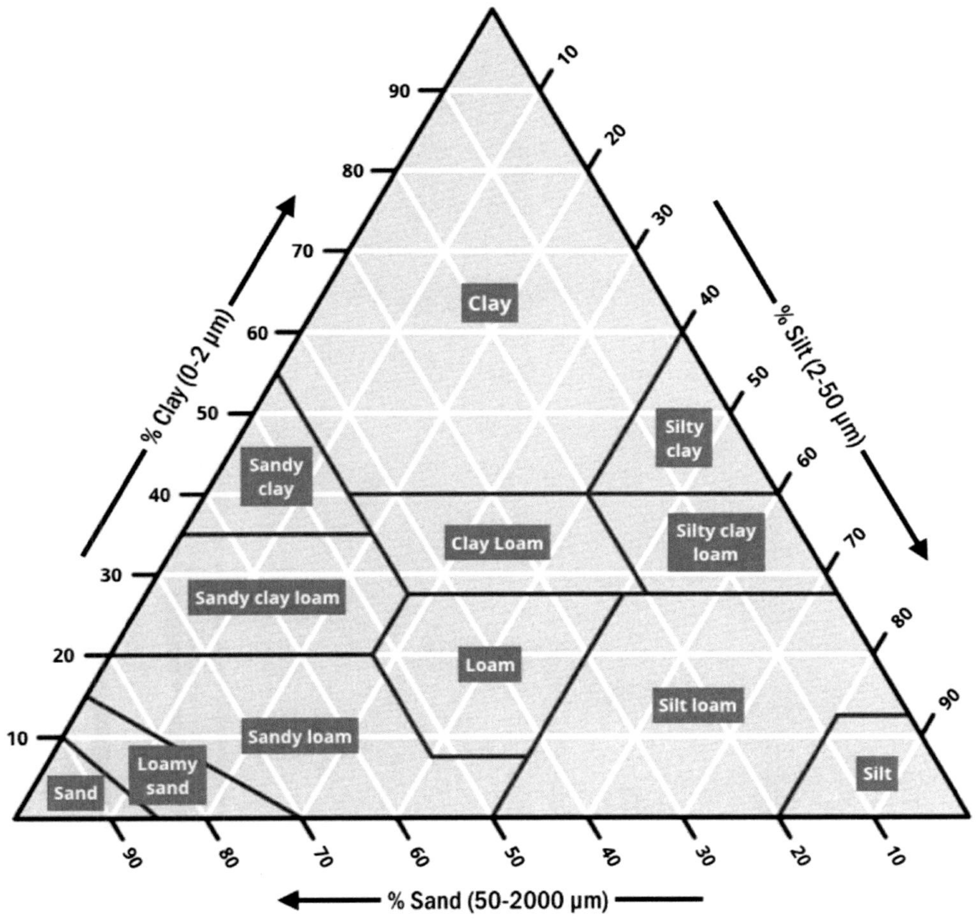

Figure 9. USDA soil texture pyramid.
The soil texture pyramid shows the ratio of particles within the soil. The middle of loam is 40% silt, 40% sand, and 20% clay.[70]

The dead organic debris that becomes stabilized, similar to finished compost, is called soil organic matter (SOM) and makes up about 5% of high-quality farm topsoil.[71] Soils have recognizable layers, called horizons, resulting from the environmental gradient between the air and bedrock (Figure 10). SOM is concentrated in the upper, or A, horizon. Even though organic matter makes up a small percentage of topsoil, it has an outsized influence on soil function.

O horizon: Organic/humus layer. Mostly organic matter like decaying leaves.

A horizon: Topsoil. Mostly minerals from parent material with some organic matter; good material for organisms and plants to live in.

B horizon: Subsoil. Rich in minerals that leached from A horizon and accumulated in this layer.

C horizon: Parent materials. The deposit at Earth's surface from which the soil developed.

R horizon: Bedrock (below C horizon/not pictured). A mass of rock (granite, basalt, quartzite, limestone, or sandstone that forms the parent material for some soils if the bedrock is close enough to the surface to weather. This is not considered soil.

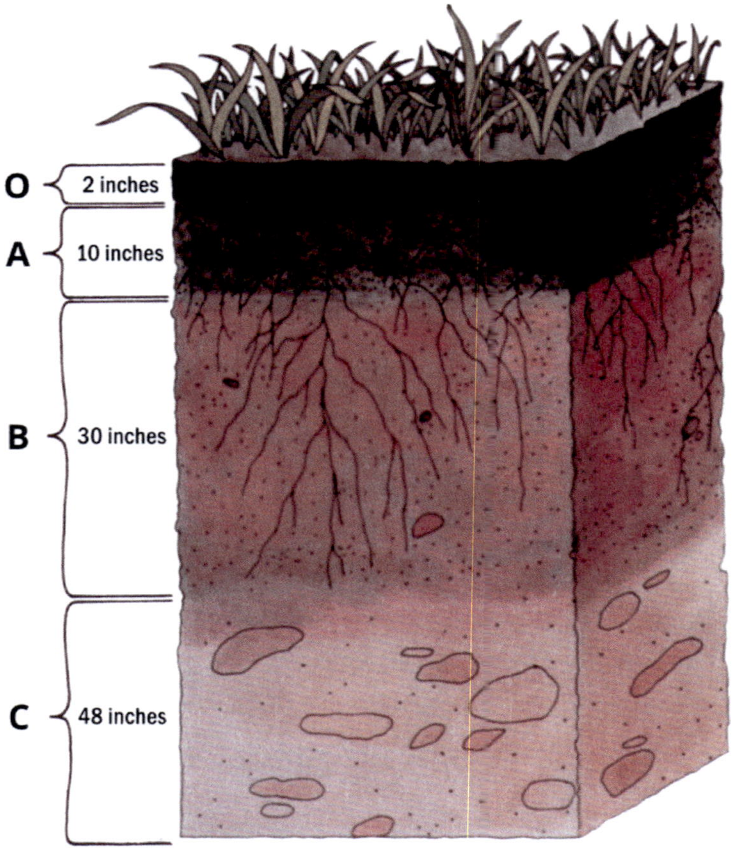

Figure 10. Soil horizons.[72]

The presence of SOM enhances crop yield potential, while reducing farming costs and risk, by helping build and maintain relationships among air, water, and nutrients. Roots and many other living things in soil need to breathe. Organic matter has sponge-like properties, creating pockets for air and water movement that can help the soil resist compaction. Soils high in SOM allow water to infiltrate the soil profile more quickly, store water well, and release it slowly between rain and irrigation episodes. Crops are therefore less likely to be drought-stressed when SOM levels are high. Soil humus ("HYOO-mis") carries negative charges that keep positively charged minerals from leaching out during heavy rains and irrigation, while still allowing those minerals to be absorbed by plants. Organic matter also acts as a buffer to keep soils from becoming too acidic or alkaline, and a more neutral pH fosters nutrient uptake by plants. Extracellular polymeric substances produced by bacteria and filamentous strands of fungi with glue-like molecules, such as glomalin, allow SOM to hold soil together, so it is less likely to erode.

The main agricultural soil is called mollisol, derived from the Latin word *mollis*, which means soft. Mollisols come from grasslands, where biological and climatic conditions are just right for rich soil formation. Prairie grasses send copious, long, and dense roots deep below ground to create pore space, promote soil aggregation, and build structure. Roots also create vertical channels for the movement of water and nutrients. Over time, as more and more root material builds up, grasslands yield large influxes of soil organic matter, which turns the soil dark brown.

Fertility and Nutrient Management

When farmers till prairie grasslands for the first time, they unlock a store of nutrients held in SOM that translates into fantastic crop yields. As the disc blades chop and churn sod, SOM is more fully exposed to air and warmth, leading to its decomposition. The breakdown of SOM is akin to what happens in a compost pile or a pile of lawn clippings. Bacteria feed off the material and release heat, carbon dioxide, nitrogen, and minerals. Having been broken down into simple chemical forms, these minerals are available for root uptake, enabling crops to thrive. The great stores of organic matter in mollisols are the original fertilizer and the catalyst for the expansion of agrarian societies that have profited from exploitation of ecological wealth built over centuries, akin to a great forest felled, but below ground and thus largely invisible to us.[73]

Most farms are losing soil organic matter through oxidation. Repeated tillage, crop residue removal, and some ways of applying synthetic fertilizers cause soil bacteria to decompose SOM faster than crops can rebuild it. In the U.S. it is estimated that about half of SOM on cropland has been lost over the past several decades.[74]

The harvest of crops removes some of the minerals that were originally in the soil. In natural scenarios, such as bison grazing on a prairie, the animals eat plants that have ingested minerals from the soil to build their leaves. The bison extract these minerals, plus energy, from the leaves to construct their bones, muscles, and other tissues. They also urinate and defecate, processes which directly return most of the nutrients prairie plants need to regrow. And eventually even the elements temporarily trapped in the bodies of the bison are returned to the soil after death.

Agronomists are aware of nutrient removal by crops, and the United States Department of Agriculture maintains a searchable database that allows a farmer to model losses from the field with each harvest.[75] One purpose of nutrient removal tables is to recommend fertilizer rates, both to compensate for what the previous crop removed and to anticipate the needs of the next crop. In industrial agricultural systems, there tend to be no qualms about purchasing mineral inputs to replace exports, although this practice is clearly unsustainable.

Most fertilizers used today are products of mining and other energy-intensive manufacturing processes. Natural gas is the chief precursor to synthesized nitrogen fertilizer. Phosphate rock is mined from a few key locations around the planet and subjected to grinding and energy-demanding chemical transformations to yield superphosphate fertilizers. Deposits from ancient sea beds are mined to yield potassium salts. Limestone, dolomite, and gypsum rocks are excavated from quarries, ground into powder and spread over acidic fields to raise soil pH toward neutral and add minerals such as calcium, magnesium, and sulfur. Other minerals may be required now and then, but those are the big ones.

If human societies were spatially organized more like a herd of bison, fertility management wouldn't be so complicated and energy demanding. People could eat, deposit their wastes, and keep their animals within the plant communities that the whole system relies on. Instead we have dissociated ourselves from productive landscapes, pay to take care of the resulting pollution in cities and feedlots (cities are functionally analogous to feedlots for people[76]), and then pay again to mine, manufacture, transport, and spread replacement minerals back to the land. This is called progress.

It would be wise to sort out how to return to land what we take from it and dispense with energy-demanding mining and processing. Returning all human waste to fields is entirely possible and was the method by which Chinese and Japanese farmers persisted for centuries.[77] In the future, we won't have the luxury of long-distance transport of food to cities and long-distance transport of wastes back to farms.[78]

Erosion Control and Soil Conservation

Farms are usually situated on the most productive soils, which means they are the quickest to develop, the deepest, and most resilient to disturbance. But "quick" in geologic time may be slow on a human scale. For example, it takes the actions of plant roots and earthworms about 100 to 500 years to produce one inch of fresh topsoil.[79] Under the cover of a forest or native grassland, erosion is similarly slow, so soils maintain a characteristic depth over long periods. However, once vegetation is cleared to plant crops, rates of soil erosion on most farms exceed soil development typically by an order of magnitude.[80] This is how agrarian societies have repeatedly failed, by farming in a manner that leads to shallower soils, which eventually causes crop yields to plummet.

Temples get buried by forests and some hundreds of year later, once the soil has been restored, the cycle may begin again.[81]

We know how to reduce rates of erosion to nearly natural levels and thereby protect agricultural soils so they can last a very long time. Soil erodes most quickly when it loses SOM and is laid bare by tillage and subject to the forces of wind and water without protective plant cover. Limiting tillage and keeping soil covered by living plants and plant residues as much as possible decreases erosion. And to protect hillsides from erosion requires maintaining perennial cover, such as pasture, or using contour tillage and terracing techniques.

The Risks of Intensification

Those rightfully concerned about the expansion of agriculture into forests and other shrinking habitats are promoting the notion of "sustainable intensification."[82] The laudable goal is to increase output per unit of land (intensification), and it recognizes the negative externalities of most current practices and aims to limit these (sustainable). Some proponents of sustainable intensification want to deliver a greener package of industrial farming technologies, such as modern seed varieties, precision fertilizer applications, and expansion of irrigation systems. Others insist on a more ecologically minded, biologically intensive approach, hoping to sidestep the industrial farming model and find a symbiotic relationship between agroecosystems and natural ecosystems. Some are critical of the notion that global agricultural output needs to increase at all given that far more food is grown than can be consumed already (it is just poorly distributed) and that models of demand for future food assume large increases in global consumer spending on meat and other luxury foods (why not improve diets? And will we really all be so much richer in the future?).

Proponent of sustainable intensification hope to avoid the errors of past agrarian societies, where population growth led to cultivation on marginal lands, which is the kind of extensification the world can't afford. Modern society can learn from cultures that farmed successfully in the same places for thousands of years, accomplished by returning removed nutrients back to the land and keeping soil covered. Still, most societies have failed to adequately protect their soil over the long run, and intensification can backfire if not done properly.

Let's unpack that last sentence, as we want to prevent an unfortunate repeat of history. The factors of production in agrarian societies are the land base, human labor, fuel, and the various tools available, which may include farm equipment, beasts of burden, and granaries. Food calories per land area are maximized by growing the seeds of domesticated grasses, such as wheat and corn, or starchy tubers, such as potatoes. As population rises, pressure mounts to increase the yield of calories for a given area, which would be intensification, and to grow food on more land, which would be extensification.

Let's imagine a scenario where extensification is impossible, so farmers must eke out more food from the same area. There are two general ways to intensify. One is to increase the yields per area of the crops being grown. The other is to shift the mix of crops and preferentially grow the ones with higher yields. Animals don't yield as many calories per acre, so in the latter case of intensification, livestock numbers would decrease as less land is allocated to feed them (and their role as a feed buffer shrinks, making the community under-insured against a poor grain harvest). As farming intensifies, farmers can harvest more calories because a greater area has come under cultivation for grains or potatoes. Some additional discussion on soil, plant, and animal biology is useful to appreciate why such intensification can be the beginning of a downward spiral.

Soil quality varies across a landscape—there are outstanding, mediocre, and inferior areas for farming. On the best soils, farmers can get high yields with modest efforts, so that is usually where farming begins. Farmers won't tend to till and plant on lower-quality soils unless absolutely necessary, because their efforts produce low returns with high risk of crop failure. Lower-quality soils tend to have shallow topsoil and are more vulnerable to erosion, so they don't have the same resilience as the best soils. A heavy rain at an inopportune time can wash away newly planted crops and the thin topsoil and make the area even less productive going forward.

Raising livestock on land of marginal quality is less risky than trying to grow a crop on it because the plants that dominate pastures are not the annual crops grown for grains, but perennials, which means they live for many years once established. Perennials keep the soil covered year-round, which helps to prevent erosion. Their roots have years to grow into deep soil layers and so have access to larger stores of water and nutrients than annuals. Grazing pressure often encourages the development of a biologically diverse field where no single plant species dominates, which usually means clovers, with their beneficial nitrogen-fixing bacteria, can become established. Like the bison on the prairie, grazing livestock redeposit most of the minerals they eat in a form that fosters pasture regrowth, and deep roots with mycorrhizal fungi can move minerals from deep soil horizons to the surface. The overall effect is an increase in SOM and enrichment of the topsoil, and farmers without synthetic fertilizers can take advantage of this effect by tilling the pasture and getting a high, albeit short-lived, boost in crop yields.

The crop-versus-livestock question for a field is not an either-or proposition, however. It may be reasonable to keep marginal-quality land in pasture most of the time and occasionally rotate the field into a crop that might do alright. For example, oats don't need soils that are as fertile or well-drained as wheat, so a field that will never yield a good wheat harvest may be fine for oats every few years. Typically, intensification leads to an increase in the proportion of fields dedicated to annual crops instead of livestock grazing, which ends up reducing soil quality across the landscape.

The example of intensification and soil decline provided above is indicative of an integrated crop and livestock system, akin to European mixed farming. However, similar dynamics apply elsewhere, even when domesticated ruminants are not included. Swidden systems (cleared using a slash-and-burn method) of sweet potatoes and taro that rotate gardens with forests would fail if forests were not allowed enough time to regrow. Growing potatoes on Andean slopes would lead to enormous soil losses if fields were not kept small and rotated across the landscape to allow native vegetation to regrow.

The key points here are that soils vary in quality, and all soils need to be managed to optimize mineral composition, organic matter, and structure if they are expected to yield indefinitely. Ruminant livestock on perennial pastures can be excellent for soil health but won't return as many calories per area as grains or potatoes. Can yields be increased while protecting the soil? Sustainable intensification is a great goal, but achieving it is complex and probably requires the full adoption by a culture that understands the challenge. In Part 4 we will learn about some of the farming schools of thought that are suited to the Great Simplification and could help achieve sustainable intensification.

Ecosystem Services on Farmland

We have been fortunate to take ecosystem services for granted until now. It is uncomfortable to worry about whether there are enough insectivorous birds and bats around to keep pest outbreaks in check. Nobody likes wondering if wetlands can buffer rivers from heavy rainfall events enough to prevent their town from flooding. Who wants to lose sleep over the question of the size of pollinator populations and whether we will have the fruits and vegetables we expect? We all can hope forests and fields will thrive, overcome the stresses of heat and disease, and can continue to drive the hydrological cycle effectively to yield enough rainfall.

As the Great Simplification unfolds, it will become more difficult to compensate for a local lack of ecosystem services, such as the provisioning of resources like food and the regulation of processes like water storage and release. To prepare for the decline of imported services and to avoid the burdensome costs of substitution with energy and technology, we can be planning to take care of ourselves by restoring ecosystem services.

Healthy agricultural landscapes can provide a wide range of ecosystem services, and we know how to foster these. A focus on soil health will improve watershed functions, such as water infiltration, storage, and quality. We can also view farms as parts of complex, diverse landscapes instead of just places to produce things we need. Landscapes tend to have various habitats related to underlying soil and geographic features. Wetlands, rocky outcrops, and river corridors, for example, can be protected and enhanced. These areas provide homes for wildlife, such as birds, bees, and bats, that are crucial partners in crop production.

In Focus: Measuring and Managing Soil Health

When farmers or agronomists take a soil sample, they typically send it to a lab for a chemistry test. The results reveal what nutrients are present in a soluble form that is readily available to plants. Deficiencies in nutrients often lead to recommendations for, and applications of, fertilizers to optimize the return on investment in crop yields. Some soil scientists have considered these kinds of chemistry tests to be incomplete and have developed methods to gauge soil health more holistically.[83] The goal is to gain insights into the biological activity of the soil and its influence on ecosystem services, including erosion prevention, carbon storage, nutrient cycling, water storage and infiltration, and flood abatement. These tests reflect a paradigm shift in what farmers should be doing to protect their soils and maintain economic benefits over the long run.[84]

Soil health metrics usually improve with: (1) decreased tillage intensity and frequency, (2) presence of living plants, plant residues, and/or mulch on fields, and (3) high crop diversity. Managing in ways that improve soil health tends to increase soil organic matter, the stability of soil aggregates, rates of water infiltration, and the rate of nutrient cycling. The biophysical processes that improve soil health occur as plants and soil organisms form symbioses where plant root exudates feed the micro- and macro-organisms in the rhizosphere, and in turn the soil biota supports nutrient uptake among plants. The soil organisms build a complex living structure, a scaffolding made from extracellular polymers, that has a variety of pore sizes and forms the habitat in which soil species thrive and interact with plant roots. This soil habitat both creates soil that is less dense with more pockets for air and water to infiltrate, and also binds particles together, which prevents erosion. Given these characteristics, it is plain to see how tillage destroys soil structure, akin to a home being hit by a tornado. As this habitat is torn apart, year after year, the pieces remain a jumbled mess. The soil lacks its scaffolding and its builders, leading to compaction and poor water and air dynamics.

Although the above relationships are well known by soil scientists, I took advantage of an opportunity to test and demonstrate soil health on some fields near where I live. The three adjacent fields had different recent histories, and I hypothesized that the soil health tests would reflect the differences. In July 2018, Shannon Cappellazzi and James Cassidy from Oregon State University visited the fields to perform tests and bring samples back to the university laboratory for analysis.

All samples were taken within a few hundred feet of each other on the same soil type to reduce differences caused by inherent variability of the soil, so we could clearly demonstrate the influence of management history on soil health. Field histories are as follows:

1. **Field 1 (C-S)** has been conventionally farmed with tillage usually occurring each year in either the spring to grow vegetables, or in the fall to sow a grain or seed crop. Winter

cover crops have been planted between summer vegetable seasons, but not with regularity. Per conventional farming practices, synthetic herbicides and fertilizers have been consistently applied.

2. **Field 2 (WP-03)** has been organically farmed since 2011 and contained a pasture from fall 2011 through spring 2017. Tillage occurred in spring 2017 and 2018 to plant annual crops, but the field had not been tilled any other times over the seven-year period. Per organic practices, some organic fertilizer was applied in 2017.
3. **Field 3 (WP02)** is organic and went into pasture along with Field 2 in October 2011. No tillage has been done since pasture establishment. No fertilizers have been applied since at least spring 2011.

We expected that soil health would rank lowest on Field 1 and highest on Field 3. With the exception of an infiltration test on Field 3, each field test was performed twice and at a different location within each field. Lab tests use soil mixed from each field location. The tables below report the results.

Table 1. Soil infiltration time and stability

	Field Tests		Lab Tests			
Field	***Infiltration Time (lower better)***	***Aggregate Stability Rating (higher better)***	***Sand***	***Silt***	***Clay***	***Wet Aggregate Stability (higher better)***
1 (C-S)	3 min 38 sec; >8 min	2; 2	10%	58%	32%	19%
2 (WP03)	2 min, 11 sec; 1 min, 43 sec	2.5; 3	15%	55%	30%	35%
3 (WP02)	30 sec; no replicate	3.5; 3.5	10%	54%	36%	62%

The field tests demonstrated clear differences. The conventional field had poor infiltration, and we even gave up waiting on one replicate that had standing water after 8 minutes. Infiltration was much faster on the organic fields, and especially the one in continuous pasture with no tillage. Soil stability tests also went as predicted, with tillage history predicting relative aggregate stability. Lab results confirmed field tests for soil stability and also showed that the soil texture was similar in all three locations (the soil texture pyramid classifies all three as silty clay loam based on percent sand/silt/clay).

Table 2. Soil chemical analyses

Field	pH	Electrical Conductivity	P (ppm)	K (ppm)	CA (ppm)	Mg (ppm)	Cation Exchange Capacity
1 (C-S)	6.05	0.068	101	237	1996	209	12.33
2 (WP03)	6.99	0.327	73	259	3158	223	18.31
3 (WP02)	7.07	0.044	54	294	2906	292	17.72

Lab tests included traditional chemical analyses. These show the pH as slightly acidic on the conventionally farmed field, which can be caused by synthetic fertilizers, but this result is still in a good agronomic range. Electrical conductivity is very low in all samples, indicating no salt build-up, which is to be expected in the wet climate of western Oregon. Higher phosphorus in the conventional field may indicate higher fertilizer use, which is common when vegetables are grown frequently. Otherwise, the nutrient levels are similar and probably reflect native soil conditions. The cation exchange capacity (CEC) is significantly higher on the organic fields, which is a measure of the soil's ability to exchange nutrients. Complex organic compounds mediate CEC. A positive feedback loop forms when soil health is high. In this loop, increasing organic matter leads to higher CEC, which helps prevent nutrient leaching, fosters nutrient exchange, and buffers pH, which all make it easier for plants to obtain nutrients, further improving soil health.

Table 3. Soil biological tests

Field	Organic Matter	C:N Ratio	Active Carbon (ppm)	CO_2 Burst 24 hours (μg CO_2-C/g dry soil/day)	CO_2 Burst 120 hours (μg CO_2-C/g dry soil/day)	NH_4-N time 0	NH_4-N time 7
1 (C-S)	4.19%	11.4	376	16.8	10.7	4.6	34
2 (WP03)	4.19%	11.4	435	20.1	13.1	2.6	80.2
3 (WP02)	5.98%	12.8	567	38.1	22.1	2.5	110.1

Striking differences show up again in the biological test results. Indicators of soil carbon, i.e., organic matter percentage, C:N ratio, and active carbon levels all trend as expected. Lab incubation tests reveal more dynamic soil biology on the organic fields, but especially on the untilled pasture. Carbon dioxide burst tests measure respiration as an indicator of the quantity of biological metabolism. Because all life is composed of carbon and nitrogen, the rate of soil carbon metabolism has been shown to correlate well with nitrogen release by microbial activity too.[85] What is interesting to note is that at time zero, soluble nitrogen is highest on the conventional field, which may reflect both lower biological activity and residual synthetic fertilizer applications. A typical soil chemistry test provides this kind of information, which is then used to suggest fertilizer rates. At first glance, the lower nitrogen levels on the organic fields at time zero would therefore suggest the need for more fertilizer than on the conventional field. However, the incubation test doesn't support such a conclusion. By following how the soil biology releases nitrogen into solution through respiration, the test reveals that more nitrogen is potentially made available to the plants on the organic fields.

Soil that holds together well and allows water to infiltrate rapidly creates an ecologically intertwined community and, in turn, supports a more resilient human society in the context of climate change and more extreme weather. Proponents of perennial cropping, whether for cutting-edge perennial grains or more traditional mixed farming with pasture rotation systems, have soil health benefits to back them up. The nitrogen cycling of such cropping methods reduces the need for external inputs while still providing good crop yields, meaning the world won't necessarily starve without synthetic fertilizers if soils are cared for properly. As Montgomery documents in the book *Growing a Revolution*, success requires adopting all three practices of conservation/regenerative agriculture: minimize tillage, keep fields covered in living plants and mulch, and maintain a diverse crop rotation.[86]

Sheep and cattle on farm fields in the early spring.

Part 4: Forward-Thinking Farming

The modern food system runs a large energy deficit and so do most farms. Our energy-sink food system is an aberration that will eventually appear as a blip in human history, and we will require our farms to run an energy surplus to avoid starvation. Studies of traditional farming systems show the surplus was typically five to ten calories returned for every calorie expended.[87] Trials in the U.S. today replicate this return.[88] Few are doing this work, and one wonders what we could achieve if more minds were invested in developing regenerative agricultural systems that aim for higher energy returns. As urban livelihoods become untenable in their current numbers and composition, this kind of work may someday become foremost in people's minds.

A food system has multiple layers and is integrated with the broader economy. But the foundation of food is still the land, how well it is treated, and what can be reliably yielded. In this part we take a tour of innovators who have applied key concepts from soil science, ecology, sustainability, and resilience thinking to the question of what kind of agriculture can conserve soil, succeed without extravagant energy use, and protect the natural environment, including the climate and other species that live with us on Earth.

The idea here is not to advocate for a particular brand or label, such as organic, but to understand what practices actually work and could be adopted by anyone who cares to farm and steward land successfully over the long run.

Agroecology

The word agroecology is a fusion of agronomy and ecology. The idea is to apply ecological concepts to agricultural design and practice.[89] An ecologist views the world through flows of energy and materials. On Earth, the energy flow starts with the sun. Plants generate material flows through photosynthesis by capturing this energy and producing biomass that animals can consume. Dead biomass and manure decompose in the soil—a living ecosystem itself—where plants can recapture nutrients and initiate another cycle of growth. With agriculture, humans are structuring an ecosystem around their own needs.

Ecologists have observed rules about what makes ecosystems function well, and these can be applied in agricultural settings, too. If we state a desired goal, we can ask what ecological design or structure could work best. Farmers want high yields relative to inputs, as well as minimal risk of crop failures. Ecologists know that to achieve these goals, the agroecosystem needs diversity.

Diversity is a critical component of agriculture at multiple scales. A pasture can have many species of forage and a variety of animals that graze on those species. A farm may have many different crops among fields. The landscape that contains the farm may have natural areas managed to promote pollination or predation services. Over time, as fields rotate in and out of crops and pasture, diversity can migrate across space.

Diversity means not simply a high number of species, but also a wide range of ecological functions. For example, perennial plants function differently than annual plants by growing over years, not just seasons, and building extensive root structures that change soil conditions. Livestock also add functional diversity on farms by increasing plant productivity and soil fertility and by affecting plant population dynamics.

How does more diversity create these desired functions? Primarily by adding to the "skill set" on the farm. Having more organisms on the farm broadens the array of skills being used. For example, grasses tend to have shallow, fibrous roots that capture near-surface moisture quickly and effectively (Figure 11). Other plants, such as white clover, have tap roots that give them access to deeper soil moisture during dry spells. Together, a grass and white clover field will be more productive than either would be alone. And episodic environmental stresses, such as drought or disease outbreak, are less likely to wipe out a diverse ecosystem because some members of the ecosystem have traits that allow them to rebound from the stress.

Natural ecosystems have evolved to optimize energy payback. The most energy- and resource-efficient land ecosystems are diverse, perennial plant communities with a healthy set of herbivores and predators. This explains why pasture helps build soil organic matter, which in turn makes is possible to grow annual crops with fewer external inputs. Many of the lessons from agroecology are practiced in organic and related systems of farming.

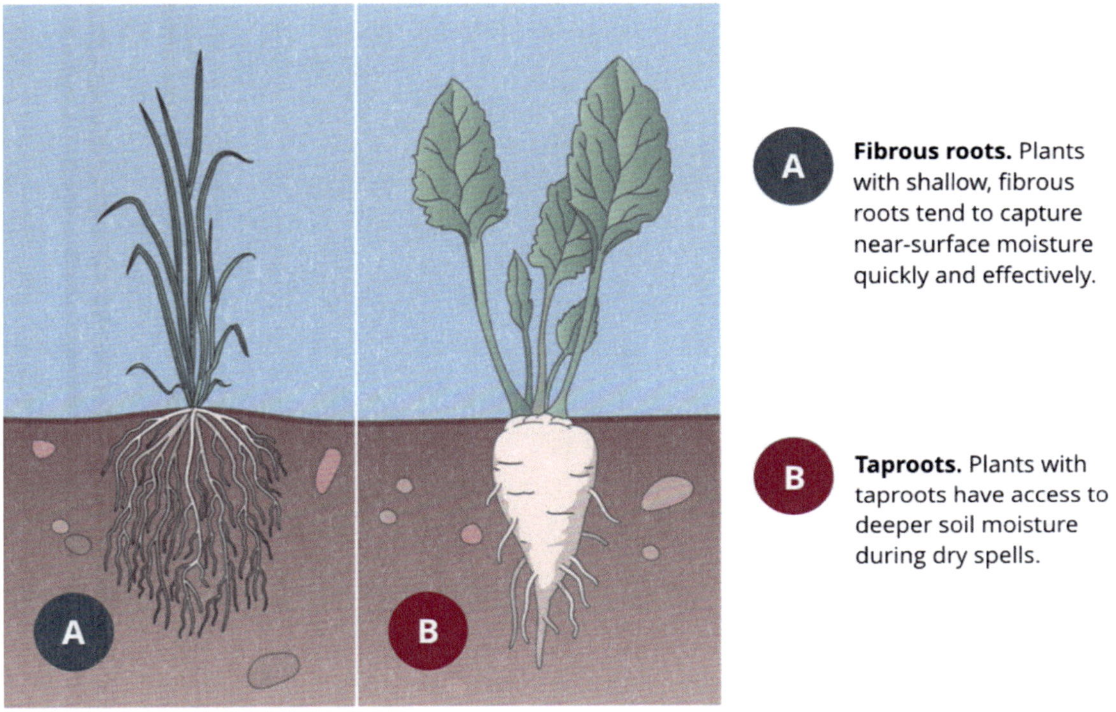

Figure 11. A plant with fibrous roots (left) and a plant with a taproot (right).[90]

Organic Farming

The term "organic" can refer to both a philosophical perspective and a legal/regulatory perspective. This distinction is captured in the USDA National Organic Standards Board definition from 1995:

> Organic agriculture is an ecological production management system that promotes and enhances biodiversity, biological cycles, and soil biological activity. It is based on minimal use of off-farm inputs and on management practices that

restore, maintain, and enhance ecological harmony.

> "Organic" is a labeling term that denotes products produced under the authority of the Organic Foods Production Act. The principal guidelines for organic production are to use materials and practices that enhance the ecological balance of natural systems and that integrate the parts of the farming system into an ecological whole.[91]

The most strident adherents to organic philosophy understand the need to reduce external inputs and farm in ways that regenerate soil health. Because organic farmers are not allowed to apply synthetic fertilizers and pesticides, they tend to adopt healthy practices such as cover cropping and crop rotations. But there can also be detrimental practices in organic farming. Perhaps the most damaging is overuse of tillage for weed control, so further development and promotion of organic methods that minimize tillage are critical.

Holistic Management

The ranching community has developed "holistic management" to achieve financial success while improving environmental conditions. The nonprofit organization, Holistic Management International, helps ranch managers understand that supporting the productivity of forage requires a deep appreciation of the ecosystems that enable forage growth.[92] The methods applied to rangeland, including thoughtful livestock movements that increase forage growth and quality, can be adapted for pastures on cropland.

The four cornerstones of holistic management are:

- Enterprise and financial planning;
- Land and animal health with planned grazing;
- Property level design; and
- Monitoring and adaptive decision making.

When managed well, grazing systems are perhaps the most sustainable way to harness biomass for human use. Aside from hunting and gathering wild plants and animals, ranching is the most appropriate method for landscapes that don't have deep, arable soils.

Biodynamic Farming

Biodynamic farming is very similar to organic farming.[93] However, biodynamic practitioners are much more likely than others to employ diverse crop and livestock rotations. While organic farmers may recognize the value of livestock integration (e.g., the soil health benefits of sowing pasture on fields, using livestock to move fertility to desired locations, and removing cover crop biomass prior to tillage), relatively few in the U.S. actually use these methods. For the biodynamic farmer, however, animal use in farming, including draft power, is a matter of quasi-religious tradition stemming from the philosophies of Rudolf Steiner. Whether one believes in the more supernatural aspects of biodynamic philosophy, the methods and the people trained in them are probably in the best position to adapt to the decline of fossil fuel energy.

Perennial Polycultures and Natural Systems Agriculture

Many of the problems of agriculture stem from the need to till the soil. The annual grains that dominate the calories in our food supply require that the ground be free of competition, which is usually achieved through tillage. No-till systems have been devised, but they usually depend on herbicides and synthetic fertilizers, which are energy-intensive to manufacture and can cause long-term declines in soil health. There are many potential advantages from getting grains from perennial plants. Soil can remain undisturbed most years, avoiding excessive use of energy for tillage and dramatically reducing erosion, pollution, and use of fertilizer.

The Land Institute of Salina, Kansas,[94] and its founder, Post Carbon Institute Fellow Wes Jackson, have been promoting a "Natural Systems Agriculture" that uses perennial seed crops instead of annuals. This is a huge undertaking as it involves domesticating wild plants. However, it is less of a stretch than most may think, as many perennial seed crops already exist, just not for food. You can go down to a local garden center and find lawn mixes, or visit a feed store and buy a pasture blend. Most of the seeds in those bags are perennial grasses, such as tall fescue, perennial ryegrass, and orchard grass. Pasture blends often contain other, non-grass perennials like red and white clover, chicory, and plantain. All of these have been bred for important characteristics to make them palatable for livestock or to make a great turf to play soccer on. If more plant breeders spent their time on perennials for edible seed traits, much progress could be made quickly.

The vision of the Land Institute is not only to have perennial grains, beans, and oil seed crops, but to grow them in mixtures, akin to a native prairie. They also envision diverse swards of perennial plants being managed by grazing. Grazing pressure can be used to encourage the growth of a desired crop and manage the timing of seed harvest. If multiple species of seeds are harvested at once, they can be separated in seed mills that sort by shape, size, and density.

Grow Biointensive Method

Farmers make decisions about what they grow and how they grow it based on socioeconomic and technical opportunities and constraints. For those with small landholdings, simple tools, and plentiful manual labor, the Grow Biointensive method is attractive.[95] Grow Biointensive was developed and promoted by Ecology Action and its founder John Jeavons in California. Their work stems from an understanding of long-term energy realities, the history of soil destruction by diverse cultures, and events that presaged the Great Simplification during the late 1960s and 1970s.

Grow Biointensive aims to use land and resources efficiently, with tools that are simple, elegant, and mostly human-powered. Soil is initially dug and loosened to create deep garden beds, which allows roots to develop freely. Compost, ideally made on the farm by harvesting cover crops and crop aftermath, is the primary input. Farmers plant a wide variety of crops with the aim of producing a complete, albeit vegetarian, diet. Plots of wheat, pinto beans, lettuce, and tomato may be found near one another. Saving seeds and growing plant starts is highly encouraged. This is the system for subsistence farmers on limited land, and for people who want to grow a large diversity of crops in their backyard.

Permaculture

Permaculture is an ethical philosophy that essentially says that to care for people is to care for the Earth. Emerging from a fundamental understanding of ecology and how energy is captured and flows through living systems, permaculture design principles provide guidance on how to modify farm landscapes. When a farm transitions through permaculture implementation, the results are strikingly different from what is practiced on most farms today but correspond to how agrarian people without access to fossil fuels have lived in the landscape. Permaculture co-creator David Holmgren gives the following twelve design principles:[96]

1. Observe and interact;
2. Catch and store energy;

3. Obtain and yield;
4. Apply self-regulation and accept feedback;
5. Use and value renewable resources and services;
6. Produce no waste;
7. Design from pattern to details;
8. Integrate rather than segregate;
9. Use small and slow solutions;
10. Use and value diversity;
11. Use edges and value the marginal; and
12. Creatively use and respond to change.

So, the first thing that we do is say, 'let's look to nature and see how nature would farm.' One of the things we know is that there is a law of return. If you take something from an area, there has to be something returned. The problem we have now in our industrial period is that we have a lot of fossil fuel and we use it for traction and for making industrial chemicals. So, not only are we disturbing the landscape, we're also poisoning it. Now, how to engage with the Earth? Some people like to garden. Some people have small farms. The larger point being that we've got to get our minds wrapped around what is necessary to end the extraction economy generally, and agriculture in particular. And so the beginning point is for people to get their hands dirty and see what's involved in, say, raising a garden. And then educate themselves on why it is and how is it that soil is as much of a non-renewable resource as oil.
– Wes Jackson

Big picture thinkers and those who like to design from first principles will be attracted to permaculture. And although most people think of permaculture as being useful for homesteads and small farms, the originators view it as a means for society to adapt to life without fossil fuels.[97] Farms in North America have been designed for machines. Historic livestock fences have mostly been removed, while field size has increased and crop diversity has withered away. This design reflects the goals of industrial, commodity-driven agriculture in which farmers have incentives to do fewer things, with fewer people, across as many acres as possible. Permaculture designers could help us imagine what farms should look like based on the ethic of caring for the Earth.

In Focus: The Three Sisters

Whereas industrial agriculture prefers the uniformity of monocropping (picture vast expanses of wheat), traditional gardening practices thrived by interspersing crops, which is often called companion planting. Perhaps the most well-known version is the Three Sisters, comprised of American corn, beans, and squash.[98]

Planting usually starts with corn in mounds spaced about 4 feet apart (Figure 12). Mounding is a simple way to prepare the ground and effectively warm the soil. Once the corn seedlings emerge, beans are sown on the edge of the mound. Beans are legumes—they partner with bacteria to fix nitrogen, making them less dependent on fertilizer. Beans tend to sprout vines that will search out the corn stalks to climb. Lastly, farmers plant squash in between the corn-and-bean mounds. The staggered timing of planting helps spread out the labor required for ground clearing. And spacing mounds widely makes it convenient to weed between them. Over time, the squash forms a dense ground cover, but the corn and beans escape smothering as their head start keeps them safely above the squash. Not only does the squash tend to outcompete potential weeds, but its prickly leaves and stems may deter animals from entering the garden patch to eat the corn and beans.

The Three Sisters demonstrate how traditional farming systems tend to apply ecological principles. Diversity is key. The three crops are from three plant families, so they are not susceptible to the same diseases. Synergies that take advantage of their distinct morphologies and physiologies make the combination not only less risky to grow, but also possibly more productive.

Beyond being beneficial to each other while growing, the Three Sisters provide complementary dietary needs. Corn is a high-yielding grain, providing a caloric staple used in breads and porridge. Beans are high in calories too, but are mainly prized for contributing protein to the diet. Squash fruits at maturity can last in storage for months, providing crucial vitamins and minerals during fall and winter. Grown and eaten together, The Three Sisters represent a nearly comprehensive garden and diet system.

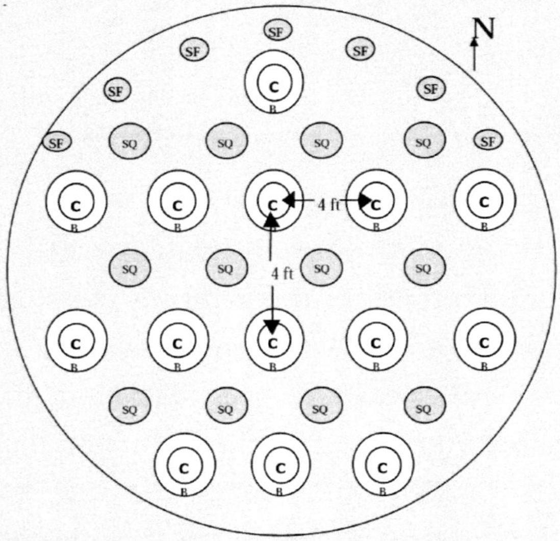

Figure 12. Circular Wampanoag garden.[99]

A swallow tail butterfly visiting the blossoms of crimson clover, an important cover crop being grown for its seed.

Part 5: Transforming the Food System

So far, we have reviewed the predicament; looked at ways of thinking that may help us solve problems in the food system; provided essential background on soils, agronomy, and the development and structure of food systems; and briefly summarized adaptive agricultural philosophies. Now the question is, how do you put this knowledge to use in your community? What strategy and tactics can guide planning and actions to change the food system in your area?

From our review of resilience science, we learned that opportunities to change arise during stress events when systems are dislodged from the basin and their state moves to thresholds (Figure 5). In a crisis people will want quick answers. If you have anticipated needs and considered some answers, then under the right circumstances you can foster appropriate responses. People are more likely to listen to you if you have credibility and social capital, so it is probably wise to be known as someone who is looking out for community interests, is able to get things done, is reasonable, and can get alone with others. And if you do need to be firm and stand your ground now and then, can you do it in ways that are sensitive to the feelings of others?

In this part of the report, we will consider some likely stress events and how they may impact the food system. To be ready to act, it helps to know a lot about where you live and expect to make an impact, and we will discuss how to conduct an assessment and build your local expertise. A dedicated group of people, perhaps a network that shares broad goals, will probably be needed to develop the models, plans, and social fabric that can grow when more people accept the need for action.

Stress Events to Watch For

As the Great Simplification unfolds we will be forced to deal with significant and sometimes turbulent changes to the food system and the broader economy. It may be helpful to have some foreknowledge (albeit imprecise) about when changes may arrive and how these can lead to the breaching of thresholds in the food system. We would like to overcome any forces that will try to pull it back into the old domain, which is unhelpful in the long run. But a system in crisis is weak and can be shifted into a more sustainable domain by those prepared to act.[100]

> *So you've probably heard of the concept of Maslow's hierarchy of needs. People need food, water, the basics, that's the bottom rung of Maslow's hierarchy of needs. If your heart-palpitating concern is that your kids don't go to bed hungry, then if your choice is to buy 10 Hot Pockets for $2.50 or three organic apples for $2.50, you're going to buy your kids Hot Pockets because your first concern is making sure your family's fed. There needs to be a level of security in order for people to care about the source of their food and make choices about the source of their food that can benefit the environment.*
> *– Kathryn Draeger*

Pressure for society to adapt could come from many angles, but a likely one is financial stress that impacts food security and political stability, perhaps leading to rational policy decisions that foster needed reforms.[101] When terms such as "more expensive" or "higher prices" are used below, relative price is more important than absolute. One key threshold is a high price of energy compared to other goods and services; such a high relative price pushes society (and the food subsystem) to reduce energy use.

Scenarios for the ripple effects of higher energy prices on the food system are outlined below. These are not based on a formal model but on my experience and knowledge of farming and food systems and their energy dependencies. However, they do correspond well to a 1976 report of an economic model for U.S. agriculture under various scenarios of energy duress.[102] While, for the most part, this report agrees with the 1976 findings for the short-term response of the agricultural system to energy stress, it also looks further to imagine the long-term implications of the Great Simplification.

A transformative shift on farms is likely to come from high-priced, persistently unaffordable, or even unavailable natural gas, and consequently expensive synthetic fertilizer. It is a wonder farms rarely use cover crops in the U.S. since legumes bring nitrogen into soil from the air. But planting a cover crop has costs. Farmers must buy seeds, use equipment and labor to plant them, and then use more resources to terminate the cover crop and incorporate it into the soil to release the nitrogen it accumulated. Farmers compare the cost of doing these activities for a cover crop to the cost of buying and applying synthetic nitrogen, which has been very cheap for the past several decades. Many farmers no longer have the equipment or knowledge to readily switch to cover crops. The natural gas and fertilizer price spike in 2007-2008 created a pulse of interest in cover crops, but cheap natural gas returned before widespread adoption.[103] Someday, perhaps soon, as shale gas fizzles,[104] the pressure to switch will be consistent.

Farms and food processors use natural gas directly for drying crops, manipulating raw crops in myriad ways (e.g., grinding, separating, blending, etc.), and cooking food to be sold in cans and other ready-to-heat packages. With more expensive natural gas, farms may favor crops that need less post-harvest drying. Fall-planted wheat, for example, may become more abundant and supplant spring-planted corn. Food processors may struggle to stay in business and those that have invested in renewable energy systems, such as biogas from processing waste, will likely have an advantage.

Diesel is critical for most modern U.S. farms as it is the power supply for tractors and combines. A sudden shortage or dramatic price rise in diesel could risk the supply of fuel to farms and lead to a decline in food production on a vast scale. Although farm equipment has become more fuel efficient since the oil scare of the 1970s, no substitutes for diesel fuel have been developed at any significant scale. The risk we face is a rapid drop in the availability of liquid fossil fuels without the time to wean farms from this key input.

Trucks and railroads don't function without diesel either. We rely on trucks and railroads to get critical inputs to farms, such as replacement parts and fertilizers, and to move crops from farms to processing and distribution centers, and finally to cities.[105]

As transportation costs rise, it will become more difficult to get vegetables, fruits, and fresh dairy foods to market. These so-called water crops are heavy, spoil quickly, and are therefore energetically costly to move. Even though the climate in California and Mexico is conducive to growing vegetables with high yields year-round, at some point it will become cheaper and more reliable to grow vegetables and fruits locally again. The nation's salad bowl is also at risk from climate extremes, rising sea level, and declining quality and availability of water, which is often pumped from deep and depleting aquifers[106] using electricity mostly generated by fossil fuels.

As fossil fuels become more expensive, electricity will cost more too. Irrigation pumps are typically electric, so higher costs will incentivize farmers to grow less water-demanding crops. Corn acres are set to decline in the western U.S. where irrigation is required. Similarly, one of the largest crops by area in the U.S. is alfalfa, and it is heavily irrigated in dry regions to serve large dairies. Dairy farms are already stressed by overproduction and low prices for milk and other products, and they are vulnerable to higher feed costs that will come with less alfalfa production and increased shipping costs. Small, pasture-based dairies, where animals can walk between forage and milking barns, will be more resilient to these changes.

A struggling economy tests thresholds in the social and political spheres. Farmers today tend to be highly specialized, and the prices of their commodities are related to the value of the dollar and the free flow of trade around the globe. As trade relationships falter, farmers may be more open to serving local markets again.

Many households in the U.S. are financially weak and have become discouraged by a lack of decent job opportunities. People may have more free time and could use practical skills that may make them more employable in sectors of the economy that need to grow, such as sustainable agriculture. Also, households today tend to be poorly versed in the arts of home economics, which would allow them to save money by buying less processed ingredients.

Across society, we have done a poor job of planning ahead and now we find ourselves in a compromised position, needing to prepare quickly to manage escalating food system crises. In the U.S. today, poverty is already rampant, but the problem of food security is related to family income and food access, not difficulties with agricultural output. As the Great Simplification unfolds and more people lose jobs, as the broader economy stumbles, and as industrial agriculture struggles to hold together, we may begin to worry about both lack of income and faltering food production. Perhaps we can avoid the sad fate of nations that have experienced famine and the resulting geopolitical unrest.[107] The next section offers some ways for communities to work toward avoiding the worst outcomes, and in doing so, find the beauty and richness of their unique places.

Strategy and Tactics

Where does someone begin whose goal is to relocalize the food system? The previous discussion of stress events and thresholds hints at the actions and opportunities that may arise sooner or later. This section dives more deeply into how to organize research and set priorities. This process can be applied at various scales, e.g., within a neighborhood, a state, or nation, but examples and potential project suggestions will focus on the intermediate scale of a community or foodshed, rather than a household or a nation. It draws from the work of many others, and is not exhaustive by any means, but hopefully serves to catalyze food system activists and make their work more effective.[108]

The broad strategy stems from recognizing the modular nature of complex systems. In a panarchy, we can live primarily in the current domain while exploring the landscape of alternatives. For example, most food may come from industrial monoculture farms, however many farmers use organic and related regenerative methods. To hasten reorganization during times of chaos, we can build knowledge and other capital, such as interpersonal relationships, that are easy to maintain and spread during the back loop of the adaptive cycle. Through our actions we may influence other parts of the hierarchy and preadapt them to more significant changes. Community conversations about cover crops, for example, may spur local conventional commodity farmers to adopt them even if they haven't bought into the notion of growing local food. And while the focus may be at the community level, households could begin developing a culture of food gardening and eating locally based on perceived future needs.

If you and others want to intercede in a system, resilience practitioners refer to an iterative process comprised of three steps: *description, assessment,* and *management.*[109] Committed groups of people in an area can use these steps to guide their work, giving it a useful structure and a sensible strategy. Each step is described in the sections below with discussion of some details that arise when considering a local or regional food system.

Description – Setting the Focal Area

The place to start is to circumscribe the boundary of the system you are going to study, and over which you may exert control or influence. This region you demarcate is called a "focal area." A focal area can be a farm or group of cooperating farmers, a local foodshed, or a set of neighboring counties. If you are concerned with supporting your local population, then you need to establish a focal area, not because you are planning for total isolation, but to be clear about where you will work to strengthen connections and build capacity. The concept of panarchy recognizes that modules connect with each other in a hierarchy. For the job of localizing the food system, you need to know about and relate to a place, make strong connections with other people and institutions, understand its position in the hierarchy, and be able to effect change within it as modules of the current food system weaken over the coming years.

Whether the plan is to start a farm, retrofit a neighborhood, or build resilient food processing capabilities, knowing the key traits, both biophysical and social, of your focal area is a critical step. For example, in an agricultural region the soil and climatic characteristics determine capabilities, constraints, and emphases. Basic questions include:

- How much of the area is considered prime farmland?
- Are significant areas rocky, hilly, and not suited to annual crops?
- What are the patterns of weather that set the schedule for agrarian life?

Because much of the diversity of local food systems has been lost, it often helps to research what people did 60-100 years ago, prior to the mass industrialization of agriculture and dominance of global trade. You may be surprised to learn that sheep were once commonly raised in the U.S. before cheap imports from New Zealand and Australia crashed the market. Or that fiber flax was a key rotation crop before synthetic fabrics took over the textile market. Perhaps turkeys were field-raised before concentrated feedlots appeared and local processing facilities shut down. In general, you are likely to find that a rich food system has been lost and that much joy and interest could come from recreating this diversity. You also have an opportunity to consider how climate change and cultural shifts could open your area to new crops.[110] The past may be some guide to a more local future, but we can also adopt new appropriate technologies, new kinds of institutional relationships, and engage with plenty of creativity and kindness to arrive at a better place—one more meaningful, fun, and sustainable.

Your place is unique. Answering the following questions can build a picture of the landscape, guide further observations, and help develop both short-term goals and a long-term vision:

- What is the population and where are people concentrated?
- Where do most people get their food?
- Is there a budding local food system that captures some share of the market?
- Does the local market emphasize certain types of foods, such as fresh vegetables?
- What do the large, commodity farms tend to grow and who are some of the key input suppliers and off-takers for their products?
- Has anyone looked at the gaps in the local food system, which are often related to processing facilities?
- Is there an opportunity to target investments to close the gaps?[111]
- Are there institutions or groups dedicated to food security and prepared to manage stresses from natural disasters and financial disruptions?
- What groups are concerned with farmland preservation, and restoration or installation of habitats on farms?
- What cultural norms, political bodies, and other social characteristics exist that could help or hinder your efforts?

Assessment – Setting Transformational Goals

In the description phase you designated and described a focal area, including agronomic possibilities, the population you are trying to serve, and the existing social fabric. Now you can build upon this knowledge to *assess the transformability* of the food system. This is when you set long-term goals and consider how parts of the food system need to change.

A commonly used assessment tool is the "diet and land model" (see the InFocus section below for a detailed example). Such a model is helpful for seeing the big picture and answering fundamental questions, such as, "What mix of regionally appropriate land and crops would it take to feed my focal area?" For this step you need to know average yields and the variance of those yields for the crops and livestock in your system. Large differences in crop yields may exist from region to region, so try to get data specific to your focal area. Also, intensive gardening systems and small farms tend to outperform large farms in yields per area, which may be important to account for.

Published diet and land models may be incomplete in the context of the Great Simplification if they assume a steady supply of exogenous energy and fertilizer.[112] When factory-produced nitrogen can't be applied reliably or is cost-prohibitive, cover crops and more complex rotations that build soil quality become crucial. For example, fields in the Midwestern U.S. will no longer be able to grow corn and soy year after year while maintaining high yields. The rotation would need to include the area and time for nitrogen-fixing cover crops, such as red and white clover, and/or livestock grazing.[113] The same principle applies in gardening systems like Grow Biointensive that aim to maintain and replenish soil organic matter and mineralizable nitrogen through crop rotation, cover cropping, and composting.

During the description process, you have become aware of the geography of your food system. Now you may want to compare what exists today with the concept of permaculture zones (Figure 13). For example, vegetables can produce significant yields in tight spaces, but they are heavy and difficult to transport, so in an energy constrained world they tend to be the focus of intensive gardening in urban and suburban areas, where they are closer to market. Moving away from population centers offers other advantages, such as the wide-open spaces needed to grow grains on a large scale or allow ruminant livestock to roam. Here you are assessing whether what exists today aligns with what is needed in an energy-constrained future, both in terms of where crops tend to be grown and how and where they are stored, processed, and sold.

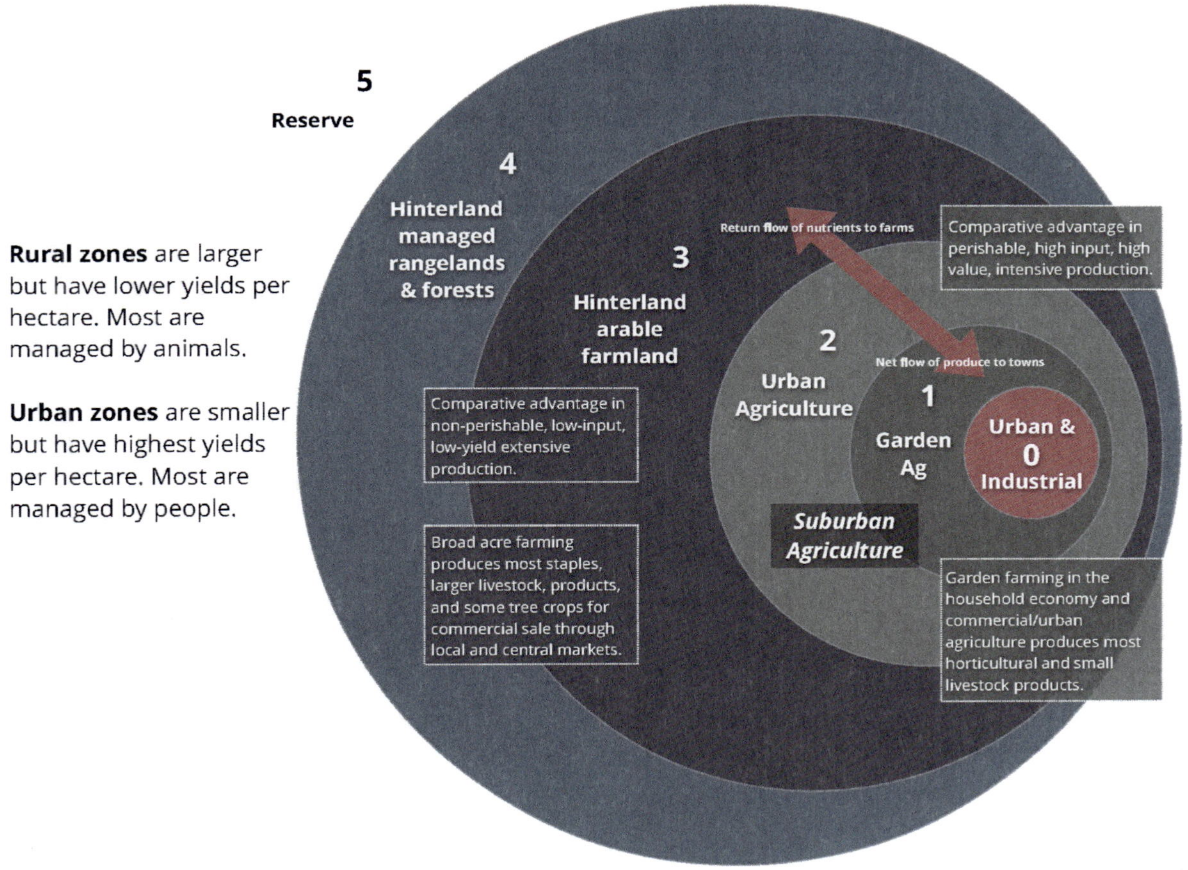

Figure 13. Bioregional agricultural zones.[114]

When all land is worked using fossil fuels, no land is set aside to feed working animals or grow biofuels for tractors and other farm equipment. The key energy input for farming with tractors and transporting goods to and from farms is liquid hydrocarbons, such as diesel and gasoline. Possible replacements for fossil stocks of these, including biodiesel, green diesel, straight vegetable oil, and ethanol, require specialized equipment to harvest, process, store, and combust. The diet and land assessment should estimate the proportion of land on a farm needed for self-reliance in energy and nitrogen while maintaining or building soil organic matter. Doing this while trying to maintain current energy consumption, i.e., substituting biofuels for fossil fuels at present day levels, is impossible.

The permaculture zone concept is important for conceptualizing the need to bring production, consumption, and recycling into proximity and thereby dramatically reduce the demand for liquid fuels. Unfortunately, few scientists and engineers are considering how to accomplish the switch to biofuels at a reasonable scale, and the answer depends on many factors, such as soil quality, climate, labor, equipment, and the mix of food crops grown. Some tasks using internal combustion engines may be replaced by electric motors, especially for lighter-duty jobs such as shallow cultivation and on-farm transport. A low-tech alternative is the time-tested use of oxen and draft horses. The land requirements to feed livestock may not be that different from feeding tractors.[115] Unlike tractors, though, livestock have multiple functions (such as fertility management and provision of meat and byproducts), can replace themselves, and can be fed from land not suited to tillage. But given our current cultural leanings toward equipment, many farmers may prefer tractors to animals.

The tractor versus livestock question (and nothing prevents farmers from selecting a mix of these) brings up an important topic, which is the characterization of labor and equipment. In general, the more a food system moves away from human labor as an input, the greater the need for equipment and exogenous energy. If energy-demanding equipment is the choice, ask whether renewable sources can be used reliably.

Richard Heinberg and Michael Bomford have presented another handy framework for determining where food production and processing should occur based on how valuable food is and how long it lasts before spoiling (Figure 14).[116] Their schema reflects how the food system should organize geographically given energy constraints. The current food system is less concerned with the distance food travels than it will need to be in the future. You may want to use this framework for both your description and assessment, looking at where different kinds of food are sourced and processed and to what degree key parts of the food system are vulnerable to liquid fuel decline. The assessment will likely crystalize priorities for localization efforts and reveal places where byproducts can be captured and repurposed.

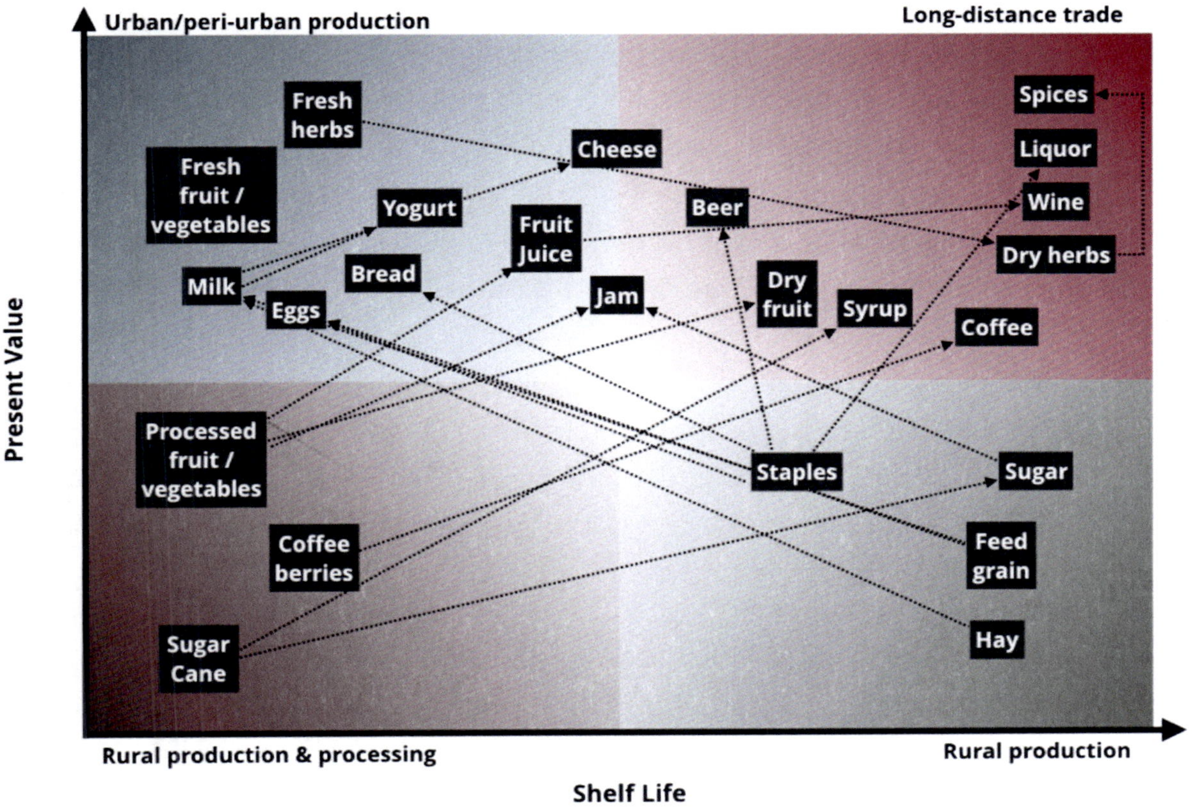

Figure 14. Geographic location of energy-constrained food production considering present value and shelf life.[117]

Per Heinberg and Bomford: "Relative position of selected food products on a plane defined by present value and shelf life. Arrows show common processes for food conversion. Long-distance hauling of food is best suited to high value, non-perishable products (red area). Urban and peri-urban agriculture is best suited to high value, perishable products (blue area). Rural production, requiring short hauls, is suited to lower-value foods for local processing (green area). Staples are usually relatively low-value, less-perishable starchy foods; urban production of staples can contribute to short-term food security during transportation crisis events, but rural production of staples is also necessary to meet long-term food needs. Processing usually increases food value and shelf life. Processing can reduce present value (e.g., drying herbs) or shelf life (e.g., making bread from grain) but not both; when present value is reduced, future value must increase through increased shelf life."

Recall the charts on energy used in the U.S. food system and notice how much occurs off the farm. The extravagant amount of energy allocated to post-farm processing supports the modern lifestyle habit of no longer adjusting eating patterns with the seasons. Recently we have seen the energy-hogging rise of frozen foods, which allow just about any food item to be eaten any time of the year. To counter this trend, traditional, artisanal, and nowadays very hip foods are being rediscovered.

Traditional methods take a seasonal surplus and transform the fresh food into something new. Hard cheeses, for example, can store surplus milk for years.

Fermentation, whether applied to vegetables like sauerkraut or meats like salami, keeps food without refrigeration or freezing.[118] Temperate-region vegetable oils tend to go rancid, and as the downsides of hydrogenation and palm oil plantations become widely known, beef tallow and lard are making a comeback. Grains can dry naturally in the field, but other foods are traditionally preserved through desiccation, including delicacies like sundried tomatoes, dates, and raisins. In the high Andes I've seen potatoes harvested and left on the ground during the dry season, occasionally stirred by feet. These potatoes, when processed in this low-tech way, form into dense starch balls that are used like dumplings in stews. Winemaking is a traditional way to preserve grapes that has never gone out of style, and many similar fermented beverages, like hard ciders and kombucha, are becoming more and more popular. Are any preserved, value-added foods being made in your area using traditional low-energy methods, and are there opportunities for growth and diversification?

"My program did experimental work with a canola farmer. The researcher purchased a mobile oil press and showed that a farmer could grow canola and use a portion of the crop as straight vegetable oil (SVO) fuel for the tractor. For example, grow 100 acres of your canola oil, use 10 acres to use for fuel for the tractor. I know there's embedded energy in tractors, but farmers are good at preserving that embodied energy, taking care of their tools. We have a solid 100-200 years of embedded energy in existing farm implements that could be maintain and used. So I think that we could have that much time worth of food, fuel, etc. for the future. Others are researching and developing solar powered tractors."
– Kathryn Draeger

Part of the assessment should investigate nutrient leakages and envision how these can be repurposed. For local food systems that require almost complete cycling of waste, a rich web of actors and processes can emerge to use resources efficiently. In practice such a web captures the output of one process to make it the input of another. Food waste becomes hog feed. Hog waste enters a biogas digester. Heat generated by the digester warms a greenhouse. Digestate residues fertilize crops... etc. Skilled jobs are needed throughout the web to envision, build, and manage relationships and tools. When the human touch is applied to such interlinked processes, the food system acts like a natural ecosystem where very little material leaks out because a diversity of species feed off one another, and sunshine moves it all. Perhaps the idea of waste needs to be discarded (pun intended) as what was formerly considered waste becomes a useful input. At the same time, we can reinforce efficient use of resources by striving for a culture that truly values food, treating it with respect and reverence.[119]

Households represent the largest consumer of energy in the food system, which means individuals and families can readily contribute by changing their behavior. This sense of agency can be an emotional salve when considering how difficult it is to change whole systems. When conducting an assessment at the household level, the question to ask is: in what ways are people who want to change supported or thwarted, either by laws, policies, norms, or lack of practical, aligned infrastructure or businesses? A cadre of advice exists to retrofit homes for energy efficiency, and similar work could apply to retrofitting modern kitchens and homes with traditional, energy-efficient features like pantries, cellars, bread boxes, etc. If people want to have backyard hens, or grow potatoes instead of a front lawn, can they do so? Are compost heaps allowed?

Thus far the kind of assessments described are useful to develop a medium- to long-term vision of an alternative food system. You may want to refer back to the *Stress Events to Watch For* section, and in a local context gauge what is termed "specified resilience," or ability to adapt to specific stresses, such as high natural gas prices. Based on what you have learned in describing your focal area and its food system, can you foresee specifically who and what may be most impacted by the stress events you are tracking? For example, let's say farmers in your area rely on a certain trade relationship; any trouble with that relationship could put the farmers in a tough position and even force them out of business.

In addition to being strategic about specific thresholds, if your work becomes well known and talked about it is likely to enhance the general adaptive capacity in your area. Complex systems are unpredictable and unanticipated events will occur, so ideally you will develop social capital and encourage widespread preparation for transformation, especially as more and more disturbance events visit your focal area.

A self and group assessment should include not only what the food system needs, but also an inventory of personal strengths and interests. Activists working on local and regional food systems ahead of, and as a response to, crisis may be a diverse group, able to cross many political, ethnic, and socioeconomic divides. Strengths could include access to financial capital, connections with aligned businesses, possession of practical farming and ranching skills, ability to navigate political dynamics and build relationships between farms and other institutions, proficiency in technical and analytical tasks, scientific knowhow and the ability to foster education and research initiatives, and expertise in cooking and regional and seasonal cuisine. Pick something that needs to be done that isn't being worked on and cheerlead those who are already doing great things. The good work you do will build credibility, and by promoting the aligned work of others, you will build the team of allies needed to make rapid system changes.

Management – Effective Tactics

You have become knowledgeable about the food system in your area, envisioned some goals for the future, and anticipated stress events that may push parts of the food system toward thresholds. Now you need to tackle specific projects, managing your time and energy effectively based on a clear-eyed understanding of priorities and available resources.

If you have a committed team, perhaps even a formal Food Policy Council,[120] it makes a lot of sense to build on the strengths of those at the table while seeking new recruits to broaden skill and interest sets. Your team may want to divide efforts into those that build local capacity in various ways, and those that are geared to responding to challenges as they arise. For example, local capacity building might include expanding the farmers market or helping start a local grain business. These are concrete projects where change is easy to measure. Responding to challenges might include development of ready-to-implement plans for scenarios you have envisioned after analyzing your local food system and anticipating how stress events will impact it. The range of possible tactics is large, and this report does not attempt a comprehensive review of the possibilities. Instead, examples are given that should spark a variety of ideas relevant to your area. Work on one component of your food system will overlap and reinforce work on other components. That's normal and desirable, as changes in subsystems need to bolster one another for the food system as a whole to transform.

Be ready for a spike in fertilizer prices. The farmers in your area may be looking for help if fertilizer prices go way up. They will need information on what can be done, such as sowing cover crops and understanding how to promote soil health, and ideally there will be someone with relevant local experience they can consult. If you are a farmer or know one you can work with, perhaps start trialing and adopting cover crops and collaborate with researchers who can measure their efficacy. At the very least, can you find local practitioners who can share widely what they have learned about cover crops and possibly other alternatives to synthetic fertilizers? Good leads may come from contacting one or more of the following:

- The United States Department of Agriculture has multiple divisions, such as the Natural Resources Conservation Service[121] (NRCS), with local offices that have staff with technical expertise;
- Land grant universities have Extension[122] offices and staff who are knowledgeable;
- Sustainable Agriculture Research and Education (SARE)[123] offers grants to field practitioners with reports on best practices and who is involved in your region; and
- Counties often have what are called Soil and Water Conservation Districts[124] that work closely with farmers.

Align local government policy. Government can play a large role in easing and enabling transition of the food system. If someone in your group has political connections and interest in public policy, they can delve into local policies and determine whether they align or hinder needed change.

Some questions to ask include: Are governments prepared for potential fuel emergencies, with plans to make sure delivery trucks and tractors are prioritized? Do land use codes make it easy or difficult to grow, process, store, and distribute local food? Have greenhouse gas reduction goals been set, and if so can they be leveraged to develop a local food system that reduces fossil fuel dependency? Does the city or county own farmland, and if so can it be devoted to local food supplies? Can policies support procurement of local foods by public institutions like schools? Can your community dedicate land in city parks and schools for gardens? Many local governments have resources to promote economic development; can these be applied to support businesses that reduce energy demand and grow local food supplies?

Incubate new farmers. Young and new farmers tend to be oriented toward local, regenerative agriculture, and we will need many more farmers in the years to come. So how can you create a place that is attractive to like-minded farmers?

Community conversations that bridge multiple stakeholders could creatively help build new farm enterprises. Those most motivated to become new farmers and ranchers usually lack the capital needed to get started and the support network to persevere. Land is costly to purchase and typically requires a 50% down payment, and conventional banks rarely make loans to new farmers. Substantial cash may also be required for equipment and operating expenses. Success in agriculture requires many things to go right all at once: favorable weather, functioning equipment, sufficient supplies of inputs ready to go, and capable people to orchestrate it all. And farmers aren't usually paid until crops have been harvested and sold, which could be a year or more after planting (or lambing or calving). In some places people with financial means are joining forces to purchase land for lease and provide operating loans to mission-aligned farmers.[125]

How do you know if a new farmer lacking experience and collateral is credit worthy? Land owners, local investors, attorneys, university extension agents, experienced farmers, economic development officers, and food activists could come together to create programs that support the next generation of regenerative agrarians. Existing programs tend to have an emphasis, likely related to community needs and the interest of founder groups, and often include training in agronomy and farm management, marketing and business, and land access and tenure.[126] Someone trained by such a program may be a safer bet to navigate the complexities of starting a farm business.

Promote the local pivot. Most of the land in your area is probably managed by large farm enterprises growing for the commodity markets. A potentially quick way to get a lot of local food production is to convince a large farm to try growing for the local market.

Farmers facing a loss of access to the global market will be more apt to localize if they can see and take advantage of local market demand.[127] You can increase local demand, enhance farm transformability, and support associated processing and distribution systems by helping shift the local culture to value seasonal cuisine. This positive feedback loop can potentially move the food system into the desired domain quite rapidly. Maybe you know farmers or landowners and can bring to their attention the topics in this report. Some farmers may be willing to change their practices and give local markets a try. Some landowners may be encouraged to lease to farmers who use organic practices and market locally.[128] Both buyers and farmers will often view these changes through the lens of their financial bottom line—they will view changes much more favorably when they can earn stable and decent returns for their efforts.

Rebuild local food processing and storage. While much emphasis is placed on growing food, don't ignore the middle ground in the food system. Processing and storage have been highly consolidated and centralized over the past fifty years, but now it is time to reverse that trend and rebuild local capacity.

During the assessment you may have become aware of local processors. Perhaps start by getting to know them, their vulnerabilities to high energy prices, and how they could adapt. Many food processors create waste that could be turned into biogas, for example.

In addition to making what exists more resilient, you will probably find many gaps. Small creameries have vanished. Granaries and flour mills have closed. Slaughterhouses and hide tanners are no more. And the warehouse districts that used to conveniently exist next to downtown commercial districts have relocated to exurbs and rural hinterlands off the interstate. It is a big thing to ask, but these will need to come back and at a scale geared to the local population, not serving half the continent or as part of a global export business.

Part of the challenge that we have to face is increasing inequality in the way that resources are distributed in our society. I'm concerned that as we face depleted availability of resources, that various members of our society will continue to control the remaining resources. The few resources that are left will have to be distributed between more and more people who are struggling for access. In terms of mitigating problems and preparing for greater resource constraint, we need to develop systems that promote equality. Systems that tax excess income and distribute income to poorer members of society. It's very dangerous when a small portion of society controls resources. I also think it's important to develop pricing systems that reflect the true cost to society of our decisions. Prices are one of the few clues that people have to work with when making decisions about what to buy. I'd like to see carbon pricing and pricing of greenhouse gases based on CO_2 equivalents.
- Michael Bomford

Improve food security for the most vulnerable. Children and young families are some of the least secure segments of the U.S. population when it comes to food access. You may be able to jump right in and work with people who could benefit from more self-reliance or increased access to healthier foods. This could take the form of school gardens[129] and farm-to-school programs,[130] or a program to improve student health through food.[131] Maybe you enjoy local foods and want to help people with restricted incomes do the same by developing ways for those on food assistance to buy at farmers markets.[132] Perhaps a vacant lot or land in a city park can be turned into a community garden.[133] These seemingly small projects may turn into crucial modules within your focal area that improve food security for all residents.

As the next recession hits and job losses bite, the underemployed may have the luxury of time, and with access to land, simple tools, and some education, they can contribute to feeding themselves (at least fresh vegetables) with gardens and small allotments.[134] A cadre of frugal master gardeners could make the leap to be farmers who feed others, and many more farmers will be needed as society descends the energy curve.

Likely the best path for a new farmer, before striking out alone, is to work with an experienced farmer. Developing education, training, and opportunities for partial self-sufficiency on even small plots of land can increase the pool of potential farm workers in your area.

Protect and enhance on-farm habitats. Ecosystem services should be strengthened to help compensate for the decline and increased costs of inputs, to protect from extreme weather, and to increase overall productivity. In most cases, the farms in your area have likely been managed to remove natural habitats, for example treating ditches with milkweeds as "unclean" and therefore sprayed with herbicide. The same agencies that promote soil health and cover crop practices are likely to have an array of helpful programs, including grant funds, available to encourage the installation of habitats on farms. Imagine beautiful, diverse native hedgerows attracting the birds and the bees, wetlands blooming and buzzing with once widespread prairie species, and riparian forests enveloping and cooling creeks that flow clear again. These are some of the opportunities you may find and be able to foster in your area by volunteering and engaging in public outreach.

Close the nutrient loop. There is usually someone who gets fired up about closing the loop on human waste. If your local sewage treatment plant is not sending biosolids to area farmland, that's something to work on. Does the company managing solid waste have some diversion and composting system for organic materials? Instead of trucking biomass away to a giant processing center, perhaps a neighborhood group can be established to manage compost heaps.

Create a local food culture. Jessica Prentice, credited with inventing the word locavore, wrote a beautiful book that aims to reconnect people to the seasonal cycles of food.[135] Although she includes recipes, her book mainly describes traditional foods from each month and how people dealt with abundance and scarcity in preindustrial times. Slow Food International, a nonprofit organization which may have a chapter in your area, is also a great place to search for information on the cultural side of sustainable and traditional food systems.[136] This is important to do because rapid change undermines people's beliefs and value systems, and many will be seeking meaning.

Most local food purchased today is probably going to home kitchens. But a local food culture needs partners with bulk buying power. Businesses and institutional buyers are leverage points to create reliable market demand for local food.[137] Your favorite restaurant in town or the cafeteria where you work could be convinced to source food from area farmers and processors and take pride in their place and relationships with local businesses. When multiple large buyers do so, it becomes part of the social fabric, and the behavior is reinforced, which is why some groups focus on farm-to-institution programs.

Learn more about renewable energy and food systems. It may be a worthy goal for most of us to know how to cook healthy meals for ourselves and our families from locally and seasonally available ingredients. Developing such culinary skills takes a level of commitment, but probably no more than what most of us spend on other hobbies or consuming media. However, after becoming proficient at seasonal cuisine, a significant challenge for cooks of the future may be sorting out what kind of stoves and ovens to use.[138] Electricity from renewable energy sources will likely be very expensive to use for cooking, and there is no clear way to swap natural gas with wood gas or biogas on a scale to supply the current pipeline infrastructure. This is a tricky problem to think through, especially in the context of a housing stock that is not designed to safely and cleanly burn wood or charcoal. As we taper and then end fossil fuel use over the coming decades, does that mean nearly all the ovens and stoves currently in use will be obsolete? Solar ovens, parabolic cookers, rocket stoves, wood gasifiers, etc. are all possibilities, but none are as convenient as what we are accustomed to having.

The kitchen dilemma is just one example of the wicked problem of facing the decline of fossil fuels. Your team could probably use someone who is excited about the long-term challenge of retrofitting infrastructure and equipment of all kinds for the Great Simplification.

In Focus: Diet and Land Modeling

In a more rural and local future, we can't rely on imported food. For this reason, people often ask how well the farmland in their home region may be able to supply food and other goods. This kind of work is often called "foodshed assessment" and is based on models that link the diet of a given population to land capacity.[139] I will review a model for where I live, Benton County, Oregon, because I know enough about the soils and cropping options here to be useful. However, for those who live elsewhere, it is most important to understand the reasoning behind the model structure, which will allow you to adapt it to your focal area by changing crops, yields, and other assumptions and inputs. The model is available for download, review, and use by others.[140]

My approach is somewhat rare in this kind of work because it accounts for the needs of a food system without fossil fuel inputs. In this way, it is like the permaculture model for Britain by Fairlie, but with the intention of making it easy to use and modify with more explicit and malleable assumptions.[141] A good reference for diet and land modeling for the United States is Peters et al., which incorporates national averages for crop yields and references to USDA-recommended dietary allowances.[142] My model is intended to understand local potential and so uses regional crop choices, and, as mentioned, incorporates demands for energy, nitrogen, and some other goods such as fiber.

The model is a spreadsheet with five linked pages. The *Inputs* page simply records the size of the population and some basic characteristics of the focal area's farmland. For this exercise, I entered the population of Benton County, which is around 91,000, and then how much land in the county is in farms. Two key metrics used in the model are acres in cropland, which is normally of high quality and can be tilled, and acres in pastureland, which may be of lesser quality and can't support annual crops reliably. All these numbers for counties can be found easily through census reports. Although more sophisticated spatial analyses are possible and could provide additional insights, my intention is to avoid overcomplicating the process and derive local food system parameters that are "good enough."[143] Perhaps a more important function than getting the numbers exactly right is using the model to think systemically and see connections.

The diet part of the model is on the second worksheet and called *Diet and Population Matrix*. This matrix places what people eat into food classes that have relatively straightforward relationships to crops and associated land. "Cereals," for example, is a food class that can consist of a variety of crops. In some places rice may be the staple, while in others it could be wheat or a basket of locally adapted grains. For each food class, the ounces consumed per person per day are entered, which is a convenient way to make the model diet relevant for one person, yet easy to extrapolate for a whole population and connect to crop yields. Table 4 below shows how the diet created for the model breaks out by food class, the resulting calories per day, and how it sums up for the population. The digestibility of the diet was checked by doubling the dry weight of the cereals and legumes to account for added water when cooking. This results in a diet of less than 5 pounds per day, which the human gut can tolerate. At about 2,800 calories per day, it is adequate energetically, and has plenty of protein, fat, fiber, and nutrient-dense foods. Only a small amount of food can be wasted to keep the population in energy balance, especially considering a future where people get around by walking and biking and perform more manual chores.

The diet includes three items that are not directly related to land area: poultry, pigs, and beer. These foods place demands on staple cereals and legumes and so act as feed buffers since they are luxury items that could be curtailed if crop yields are low. Hogs are also especially good at eating food waste, poor-quality cereals, and seed meal coproducts, and so when raised in their evolved context (i.e., not the modern CAFO system), they can increase the food supply on small amounts of land. Certain complexities are overlooked in this model. For example, beef tallow and pork lard are not contributing to calories or reducing the vegetable oil demand. Also, dairy is modeled as fresh milk, and none is converted to butter, cheese, or yogurt. I assume including these details would be a lot of work while not changing the results significantly.

Table 4. Diet and population matrix

Item	Consumption/ person/ day (ozs)	kcals/ oz	kcals/ person/ day	Consumption/ population/ year (lbs)	kcals/ lb
Directly Converted to Land					
Cereals	8.0	95	760	16,607,500	1,520
Legumes	3.0	100	300	6,227,813	1,600
Oils (vegetable)	1.5	250	375	3,113,906	4,000
Vegetables	10.0	9	90	20,759,375	144
Fruits	6.0	12	72	12,455,625	192
Nuts	2.0	160	320	4,151,875	2,560
Sugar & Honey	2.0	112	224	4,151,875	1,792
Milk	21.0	16	342	43,594,688	261
Beef and Lamb (grass fed)	2.5	73	183	5,189,844	1,168
Wine	2.0	20	40	4,151,875	320
Indirectly Converted to Land (feed buffer)					
Poultry (eggs)	0.4	42	17	856,324	672
Pigs (pork and bacon)	0.3	100	34	700,629	1,600
Beer	4.0	13	52	8,303,750	208
Total			**2,809**		

The third worksheet, the *Land Requirement Matrix* (Table 5 below), is the largest and most detailed, and does the conversion from the population-level diet to the land required. This step requires knowing about a local environment and what crops are most suitable. Yields per acre for the crops are entered, and the spreadsheet simply divides the pounds of food needed by the yield per acre to derive acres to grow. In the Benton County example, about 6 million pounds of dry beans divided by 1,500 pounds per acre requires around 4,000 acres to sow and harvest. Yield statistics are for organic farming methods because the model assumes limited availability of synthetic fertilizers and pesticides in the future. While organic crop yields shouldn't differ from today's conventional yields when grown by skilled farmers,[144] I entered conservative yield estimates.

Although the math for staple crops is straightforward, the math for the livestock portion of the diet is not. Complexities for cattle, sheep, swine, and poultry include the following calculations and assumptions:

1. Some proportion of beef is a byproduct of dairy animals, such as cull cows and calves not kept as replacement stock. The size of the dairy herd is a function of the demand for milk and the assumed yield per cow. Modern yield information for cows primarily comes from large confinement systems that won't be viable in an energy-constrained future. Hardier breeds of dairy cattle that can walk to their forage will likely have lower milk return per animal each year, but also live longer than today's breeds. A large dairy herd with slow turnover results in many animals used for beef.
2. While meat supply is directly related to growing beef calves and lambs, the forage/land needs of the livestock system also includes support for breeding stock.
3. Mature ruminants such as cows and ewes don't need to graze on the most productive soils all the time, and so some of land for livestock does not overlap with the and that grows crops. The productivity of pastures on higher- and lower-quality land needs to be estimated, as well as the proportion of forage allocated to each land type in order to estimate the number of acres required. The resulting amount of cropland used by livestock has implications for crop rotations and soil health.
4. Ruminant livestock can feed on crop aftermath and cover crops to some extent, and this must be estimated, which reduces the exclusive area needed to support them.
5. For monogastric livestock, such as pigs and chickens, there is a feed conversion ratio between what they ingest and how much meat or eggs are produced. The parent stock must be fed as well. I modeled a local feed mix to incorporate into the cropping acres with this basic recipe: a third cereals, a third legumes, and a third seed meal as a coproduct of oil seed crops grown for biofuels.
6. Hogs are especially good at eating food waste, so some portion of their feed mix can be offset.

Detailed assumptions and further explanations for the calculations are provided in notes within the spreadsheet and won't be reviewed here. Rather than reviewing numbers that are estimates, the useful concept revealed by the model is the relationship among components of the integrated food and energy system. For example, the nominally non-food crops (oil seeds for biofuels) have a coproduct of seed meal, and the spreadsheet checks to see that enough seed meal is available to meet the feed blend needs of the hogs and poultry.

Imagine the year is 2050, not that far into the future. By then, according to reports from the Intergovernmental Panel on Climate Change (IPCC), we should have discontinued the use of fossil fuels.[145] The technology I believe has the best chance to power farm equipment is plant-based oil, either used directly, or converted into biodiesel or green diesel. I selected area-appropriate crops (canola and camelina), obtained their seed yields from the literature, and estimated how much oil can be pressed per pound of seed, which gives a calculation of oil production per acre. Estimated tractor fuel demand, then, determines the number of acres that need to be sown in oil crops. Fuel demand is derived by looking at the mix of crops in the model, applying a gallon-per-acre intensity, and then multiplying by the number of acres for each crop class. For example, growing a tractor-intensive annual crop like vegetables may take 6 gallons per acre per year, whereas running livestock over perennial pasture may only require one. To get total acres to sow in oil seed crops, I multiplied this on-farm fuel demand number by three to provide some liquid fuel off-farm, and then divided by gallons of seed oil produced per acre. Reducing the fuel intensity of farming is certainly possible, especially when incorporating the best practices of regenerative agriculture, such as minimal tillage and use of cover crops and mulches to suppress weeds.

Two other non-food crops are simple to estimate. Flax and hemp are potential local fiber crops. The Benton County model estimates per capita annual production of 12 pounds per year. Seeds and nursery stock land needs are simply estimated at 1% of the area farmed in annuals and orchards.

Green manure crop area is more significant and stems from some key assumptions. The farming system being modeled expects to meet its nitrogen needs from natural processes to avoid any dependence on natural gas feedstocks. Renewable electricity, perhaps from seasonally abundant hydropower, during the summer from photovoltaics, or on windy days from turbines, can be used to generate the hydrogen needed for ammonia synthesis.[146] While that technology may be developed to support agriculture, the scale of the future industry and competing demands are unknown. What is well known and technologically feasible is using legumes in rotation to build nitrogen and carbon stores that are decomposed by soil organisms and then available to plants. What I estimated is how much additional area is required to allow the nitrogen fixing plants to develop fully. Ideally, a legume-dominated cover crop mix is sown and has enough good weather to mature and be incorporated into the soil as a "green manure" with plenty of time left to plant a subsequent crop that will enter the food supply. However, weather can be fickle, farms are complex to manage, and not everything goes as planned. So, I assumed that 25% of the time an annually cropped field can't be planted in both a cover crop and a food crop within the same year. The short growing seasons of high latitudes will require more area to be set aside for cover crop development, while a farmer in the tropics with year-round good weather may always be able to grow a successful cover crop. The best cover crop farmers perform inter-cropping, where, for example, a clover is sown along with a grain crop so that the cover crop can take off after grain harvest with no additional soil disturbance or tractor passes. Knowledge and equipment that support effective integration of nitrogen fixing cover crops are going to be in very high demand during the Great Simplification. As perennial pasture is tilled under, it is akin to a cover crop, and so I credited a 25% portion of the pasture acres on cropland to supply the nitrogen needs of annual crops. Another assumption is that human and animal wastes are brought back to farms and spread onto perennial and permanent crops. This brings nitrogen back to the fields along with other nutrients that are not available from the atmosphere, such as phosphorus. If people are uneasy about applying biosolids to cropland, they could be applied only to pasture. Vegetable acres are assumed to get their nitrogen from seed meal and cover crops.

The acres needed to support the population of Benton County are shown in Table 5 below. A notable model result is the acres per person at 0.42, which seems reasonable for a productive landscape. I'm also encouraged that only 6% of the area is given to biofuel crops, even with substantial off-farm exports. My worry is that this result is too optimistic. Very little research and few experiments are being done to see how farms can be weaned from fossil fuels,[147] which is amazing considering how quickly we need to do that. The model does not assume any substitution between liquid fuels and electricity, for example, to perform on-farm tasks. Certainly, much light-duty work could be performed by electric motors with on-board battery storage, although given power demands, the prospects of battery-electric systems for tillage and harvest activities are dim.

The size of the feed buffer is the land attributed to poultry, pigs, and beer, which is about 1,700 acres. This is compared to over 10,000 acres devoted to direct consumption of cereals, legumes, and biofuel oils. Given the inherent variability of crop yields, I'm concerned that this buffer is rather thin.

Before concluding that this model is useful or realistic, it is important to evaluate whether it meets some basic agroecological standards. The next worksheet, *Agroecology Check*, classifies each crop class by longevity and plant family. Annual crops have the highest turnover, and farmers usually disturb the soil to grow them. Perennials last multiple years and so allow time for regeneration of soil structure. A field can rotate between annual and perennial crops. Permanent crops include orchards and vineyards, which may last decades and so are excluded from rotation plans. By classifying the dominant crops and summing how many acres they occupy each year, we can see the proportion of fields in annuals, perennials, and permanent crops. Diversity is also a key attribute of farming systems that promote soil health and reduce risk. The plant family is a good taxonomic category to see if the system is diverse because basic biological structures, functions, and physiologies are often conserved within families but differ between them. For example, all members of the legume family, *Fabaceae*, will fix nitrogen. Species in the grass family, *Poaceae*, tend to have fibrous root systems. Plants in the mustard family, *Brassicaceae*, usually have tap roots and produce phytochemicals called glucosinolates that affect human, animal, and soil health. When growing annual crops on a given field, it is a good practice to avoid growing plants from the same family in consecutive years to reduce the risk of a crop failure from disease.

Table 5. Land requirement matrix

Item	Consumption/ population/ year (lbs)	Yield/ acre (lbs)	Acres needed / year	kcal/acre	Acres as % Total
Directly Converted to Land					
Cereals	16,607,500	4,000	4,152	6,080,000	11%
Legumes	6,227,813	1,500	4,152	2,400,000	11%
Oils (vegetable)	3,113,906	798	3,902	3,192,000	10%
Vegetables	20,759,375	30,000	692	4,320,000	2%
Fruits	12,455,625	25,000	498	4,800,000	1%
Nuts	4,151,875	2,000	2,076	5,120,000	5%
Sugar	4,151,875	6,000	554	10,752,000	1%
Milk	43,594,688	11,208	3,890	2,923,039	10%
Beef and Lamb (grass fed)	5,189,844	491	10,580	572,963	28%
Wine	4,151,875	5,600	741	1,792,000	2%
Indirectly Converted to Land (feed buffer)					
Poultry (eggs)	856,324	550	1,043	369,600	3%
Pigs (pork and bacon)	700,629	550	420	880,000	1%
Beer	8,303,750	16,000	259	3,328,000	1%
Non-Food					
Fiber crops (e.g., hemp, flax)	1,100,247	3,600	306	N/A	1%
Biofuel crops (e.g., canola, camelina)	1,355,260	640	2,118	N/A	6%
Seed and Nursery Production	N/A	N/A	209	N/A	1%
Green manure crops (N fixation)	N/A	N/A	2,674	N/A	7%
Total			**38,266**		
Acres Per Person			**0.42**		

The *Agroecology Check* (Table 6) is shown below. Notice the preponderance of either *Poaceae* or *Fabaceae* each year. This is to be expected as grains and legumes are dietary staples. We'd like to see enough crops in other plant families in the rotation mix to give fields a break from those two families. Also of interest is the proportion of cropland acres in annuals or perennials. The cropland fields that rotate between annuals and perennials are in annuals 64% of the time. Just looking at the annual part of the rotation, both *Poaceae* and *Fabaceae* crops are grown every third year, showing that other families, especially *Brassicaceae*, add needed diversity. Given the rotation into perennial pastures, the average time between a cereal grain or a dry bean crop is reduced to every 5 years (e.g., 30% times 64% = 19%), which is a safe amount of time between plantings of the same crop.

Table 6. Agroecology check

Item	Acres needed/ year	Type	Plant Family (Most Common)
Directly Converted to Land			
Cereals	4,152	Annual	Poaceae
Legumes	4,152	Annual	Fabaceae
Oils (vegetable)	3,902	Permanent	Oleaceae
Vegetables	692	Annual	Mix
Fruits	498	Permanent	Rosaceae
Nuts	2,076	Permanent	Betulaceae
Sugar	554	Annual	Chenopodiaceae
Milk	3,890	Perennial	Mix
Beef and Lamb (grass fed)	10,580	Perennial	Mix
Wine	741	Permanent	Vitaceae
Indirectly Converted to Land (feed buffer)			
Poultry (eggs)	1,043	Annual	Poaceae/Fabaceae
Pigs (pork and bacon)	420	Annual	Poaceae/Fabaceae
Beer	259	Annual	Poaceae
Non-Food			
Fiber crops (e.g., hemp, flax)	306	Annual	Linaceae
Biofuel crops (e.g., canola, camelina)	2,118	Annual	Brassicaceae
Seed and Nursery Production	209	Annual	Mix
Green manure crops (N fixation)	2,674	Annual	Fabaceae
Total	**38,266**		

Table 7. Agroecology check metrics

Metric	%
% Annual Crops	43%
% Perennial Crops (includes non-tillable area)	38%
% Permanent Crops	19%
% Annuals in Rotated Cropland Acres	64%
% Perennials in Rotated Cropland Acres	36%
% Cereals (of all annuals)	31%
% Legumes (of all annuals)	29%

My main concern when looking at the farming system implied by the model is how soil health will fa r given the dominance of annuals on cropland. The system modeled has about 2 years in annuals for every one year in perennials (which in practice would be more like 6 years in annuals and 3 n perennials), posing a risk on soil health due to tillage. Reduced-tillage annual cropping systems that avoid herbicides and synthetic fertilizers do exist but have not been widely implemented. Building up organic matter, interspersing cover crops into food crops, diversifying rotations, and using tools that minimize soil disturbance while still promoting seed germination and crop establishment are essential practices to learn about and adopt widely. [148]

The last part of the model, *Sufficiency Potential* (Table 8 below), compares the land resources needed to support the population with the land that exists in the area. Nearly all the crops in the model need high-quality land ("cropland") to grow well. The two exceptions are pasture for cattle and sheep, and wine grapes, which I allocated half to high-quality land and half to marginal land ("non-cropland"). The table below shows the resulting acres needed of both cropland and non-cropland for the model. I then went back to the *Inputs* worksheet and pulled in two data points, cropland acres (assumed to be mostly high-quality land) and pastureland acres (assumed to be mostly poor- to marginal-quality land), from the county records. For cropland acres, I made an adjustment that needs explanation. The use of synthetic fertilizers and heavy farm equipment has allowed farmers to plant crops on marginal-quality soils. Agricultural census data tell us what farmers are doing today, not what would be done if soil health were a consideration and if cheap fertilizers and fuel supplies were lacking. A detailed, bottom-up, geographic analysis using soil survey information could probably reveal the extent of over-planting on marginal land that is happening.[149] To keep things simple, and realizing this exercise is about thinking in systems more so than making perfect calculations, I estimated that about 10% of current cropland acres would be unavailable in a future with fewer resources and therefore reduced the acres from the census accordingly ("Adjusted cropland available in area"). These acres haven't disappeared but are probably better suited to other land cover types, such as wet prairie, permanent pasture, agroforestry, riparian woodland, or upland oak savannah.

Table 8. Sufficiency potential

Land Type	Acres
Cropland acres needed per year	32,605
Non-cropland acres needed per year	5,661
Adjusted cropland acres available in area	61,368
Non-cropland acres available in area	29,754
Cropland Surplus (or Deficit)	**28,763**
Non-cropland Surplus (or Deficit)	**24,093**

Subtracting what is needed from what is available for cropland and non-cropland area tells us whether the local food system would have a surplus or deficit. For Benton County, a substantial surplus is available. Presumably this rural area, rich in fertile land, could export food, fiber, and fuel to cities, ideally with the caveat that biosolids and composted waste return to the farms and support an influx of population as cities decline.

Farms are one key part of the food system. This model analyzes practices on farms and the material requirements of a locally self-reliant agricultural system. Not covered are the wider boundary components of the food system, such as transportation, storage, food processing, retailing, cooking, waste recapture and return, and equipment manufacture and repair. Developing a fossil-fuel-free energy system to drive the other layers of the food system and broader economy is going to be a challenge in the timeframe available to us. Liquid fuels were modeled for on-farm and some off-farm use, which may partly support transportation needs. However, only about 1.3 gallons per person are supplied each year. Wood products will be a major contributor to our energy future, and to some extent this may be a byproduct of food production, such as orchard pruning residues. Pasturelands can be integrated with silviculture, and edge space on farms can support coppiced hedgerows. Wood-based energy systems can feed back into soil health through biochar products.

A very worthy study, which is beyond the scope if this report, would be a deep dive into the potential for more local energy systems to support broader needs. Without a doubt, needs will have to be reconsidered and reprioritized, as local and renewable energy systems will only support a much smaller, leaner economy compared to today.

Finding Meaning and the Inspiration to Act

When rebuilding a local food system, it can be overwhelming to consider how far we are from achieving long-term goals. But historic perspectives and strategies offer hope. During the 1940s most food eaten in America was produced locally, and farms tended to be much smaller and with fewer external inputs compared to today (Figure 15). That was quite a few decades ago, but within one human lifespan. If 5% of farmland per year in your region transitions from serving the global commodities trade to serving local populations with regenerative soil practices, then in 14 years half the land would have made the shift. For anyone doing this work, the point is to appreciate small beginnings and seek to build momentum over a lifetime. It is also possible for an incredibly rapid transformation to occur when the need is apparent, such as the switch to a war-time economy, so it's well worthwhile to prepare for when the time is right.[150]

Most people are not politically extreme or dogmatically ideological. However, being politically engaged will mean encountering the predominant mindset of American culture: the intertwined myths of progress, technological optimism, and growth. How do you navigate social situations when denial over the end of growth is so apparent? It is probably best to be polite and respectful in nearly any situation, but don't hesitate to point out dissonance between proclamations, actions, and outcomes.[151] Help people get beyond the delusion that our predicament will be solved by the market and technology, and that we can grow our consumptive lifestyle far into the future. We need many bright and engaged people to put their creative force behind managing a transition that is hard to imagine.

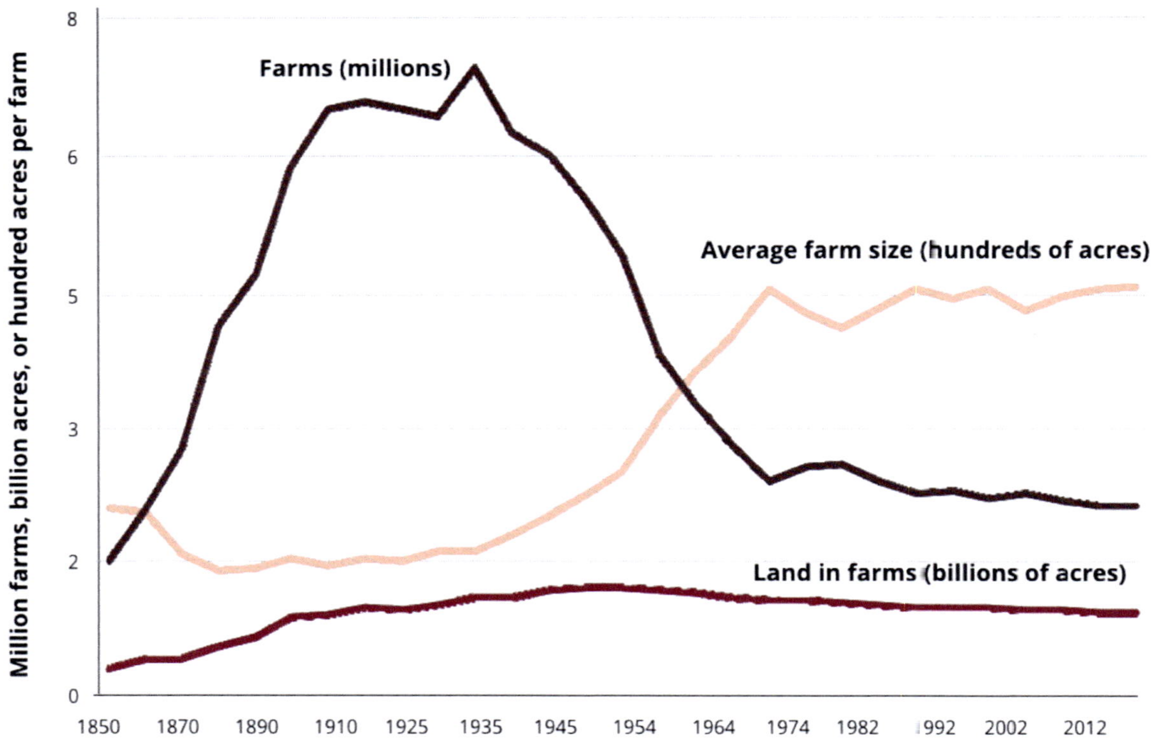

Figure 15. Farms, land in farms, and average acres per farm, 1950-2016.[152]

The food system of today has evolved by following financial incentives (admittedly short-term and supported by many policies that make no long-term sense). Rebuilding local and regional food systems in the near term, while labor is still expensive and fossil fuels are cheap, is not necessarily a highly profitable business opportunity. Those who invest in alternatives may be out of sync with business trends and risk financial losses. The fortuitously wealthy can weather the costs of a new and risky business, or devote personal time to working without pay, but most people need to make a living here and now. Community volunteers and activists who can invest time hopefully will find such efforts as their own rewards. Perhaps there is little harm, and a lot to gain, in learning to live more thrifty, skilled, and resilient lives.

Very likely, financial, regulatory, or other considerations will require the use of nonrenewable resources and prevent practices like full nutrient cycling back to farms from being realized immediately. Honestly, being sustainable is nearly impossible without the surrounding system in full support. Be mindful of shortcomings so they can be addressed later, but don't let imperfections impede progress. Like climate change and the energy system, food system reform is a collective problem, so work hard on social and political fronts and forgive yourself for personal inadequacies.[153]

Those supporting local foods with their dollars are usually buying because they perceive values beyond immediate nutritional requirements. While this may seem elitist, it is also a situation to be thankful for. Their willingness to support local farms and businesses provides a base of infrastructure and cultural awareness that will someday me much more than a luxury.

Places that successfully transition away from dependence on global commodities may enjoy the kind of food system that protects soil, embraces biodiversity, cleans watersheds, provides local employment, reconnects people to the land, and enhances community security. Growing and eating local foods builds a sense of place and a lifestyle more in tune with seasonal rhythms. If internet shopping and overnight shipping are not a big part of the future food scene, we may expect more social interaction and cohesion than we have today, which would be a great countertrend to the isolation and loneliness many suffer. And the existential angst many people feel could dissipate as they develop new competencies, undertake more meaningful work, and play a role they believe in.

Good food is more than a means of subsistence; it is ultimately a deep cultural trait. As the Great Simplification unfolds, we have the opportunity to create new and beautiful ways to live on Earth. The prodigious energy contained in fossil hydrocarbons was an amazing gift, but it came with a hefty price, including real losses of regional self-reliance. Some worry that in losing access to cheap energy the worst aspects of the past, such as xenophobic tribalism, will resurface. Navigating energy descent will likely require that we take the best of liberal world views—like openness to ideas, enthusiasm for change, and tolerance of differences—and at the same time be deeply conservative—valuing interpersonal relationships, demanding respect and civility when those are under threat, and rediscovering a land ethic that resets social norms to help us restore and protect the places we love.

Appendix

Other Voices

During July and August 2018, research assistants Miranda Edwards and Ayana Ito interviewed a few thought leaders and practitioners to get their perspectives on the topics in this report. Below are their brief bios and selected responses to questions posed.

Michael Bomford, PhD, is currently on the faculty at Kwantlen Polytechnic University in British Colombia. His research and teaching career has focused on organic systems, energy and farming, and ecological pest management.

Kathyrn Draeger, PhD, lives in Minnesota where she helps manage her family farm while working as the Statewide Director for the University of Minnesota's Regional Sustainable Development Partnerships. Current projects include promoting local food systems that recognize existing assets such as rural grocery stores and wholesale distribution routes, as well as work on passive solar greenhouses, known in Minnesota as Deep Winter Greenhouses.

Mike Eaton worked on land conservation in California for The Nature Conservancy and other organizations. He recently left a small farm south of Sacramento, where for 15 years he grew fruits and vegetables for food banks, restaurants, and local subscribers. He now lives in Berkeley, California.

Kelley Eicher is a Wisconsin cash crop and beef farmer, focused on farming practices that involve natural systems for pest control, soil and water conservation, and nutrient retention and recycling. He is also dedicated to the humane treatment of livestock and has a BS from the University of Minnesota.

Wes Jackson, PhD, is founder and President Emeritus of The Land Institute in Salina, Kansas. He is the author of several books and has been recognized as a leading international figure on the topic of sustainable agriculture.

Kenneth Mulder, PhD, has a degree in ecological economics, teaches math at Green Mountain College, Vermont, and between 2007-2013 was manager of the university farm where he compared the use of human, animal, and engine power to grow vegetables.

What do you think is essential for everyone to know about our food system, given the realities of resource scarcity in the future?

"I would say know your area, know the foods that grow easily and comfortably and nutritionally where we're at, know how to prepare them, cook them, and preserve them, and preserve them in multiple ways." – Kathryn Draeger

"We've got to get used to the idea that we have to end energy and material growth. We've got to start living within our limits, and that will solve a lot of problems. But that will be the greatest transition and transformation that humanity will have ever made. Since we started agriculture some 10,000 years ago, we have gone after various pools of energy-rich carbon. The first pool being grain agriculture, where we had the young pulverized coal of the soil. And then 5,000 years ago, the forests were used to smelt the ore for the Bronze Age and Iron Age. And then there's coal and natural gas in our time. During that time, the population has increased. We're now around 7.5 billion, or headed that way, and we're dependent on fossil fuels. Without the Haber-Bosch process, 40% of us wouldn't be here. We've got to address that. That's an energy-intensive approach. There's also an information-intensive approach. That would be the biological nitrogen fixation: the interaction between the bacteria and the legume roots using 21 enzymes. That's not an energy-intensive, but an information-intensive approach. To move from energy-intensive to information-intensive is what's going to be required of us." -- Wes Jackson

"Farmers are smart, and they can solve these problems, but they need incentive in order to start doing this. They need the playing field to be level, and that's where economics has to come in. For example, it would be nice to be raising older, hardier breeds of livestock, but you're going to put twice as much food in and get half as much meat out. Farmers that go that route are competing with meat supplies that are getting steadily cheaper and cheaper. The agribusiness is pretty darn smart, and they're responding to incentives. Right now, that's produce as much meat as possible. Imagine making it produce as much meat with as few resources and little environmental damage as possible and making there be a market incentive." – Kenneth Mulder

"A lack of awareness of the importance of soil and building SOM and sequestering carbon in our soils. For most of our economy, the relationship between burning fossil fuels and greenhouse gases is rock solid. In ag, that's not the case. In ag, our greenhouse gas emissions come from the way we manage our soil and livestock. I think it's important that people recognize the importance of soil and livestock and the problematic assumptions in systems that rely on soil or ruminants that are fed grain from farms and that excrete a lot of waste." – Michael Bomford

What skill sets do you think people should develop to prepare for the Great Simplification, and are there things you are doing in your own life?

"Learn how to work with bodies and tools. Work hard for extended periods of time. We have a generation that didn't have to work growing up, with no knowledge of the skills or the joy of it." – Mike Eaton

"The ability to be happy with less. Force yourself to not bring your cellphone places, not have your music, eat simple foods. That ability to simplify voluntarily and realize how enjoyable it can be. We are pretty hardwired to do it. In ecological economics there's the 7-point theory that says that you come back to a certain setting of happiness after momentarily being on a high. If you have a major setback, similarly, in a short amount of time, you come back. You need that knowledge to know that you can get rid of things you don't need. There will be a short adaptation period, but they're hardwired to adapt." – Kenneth Mulder

"Closing the nutrient cycle (returning all animal wastes to the soil) is important on both macro and micro scales now and will become critical when growing food locally becomes a necessity. As we move toward a warmer, post-carbon world, synthetic chemicals and bulk imported manures will be unavailable or unaffordable and maintaining soil health will be paramount. The simple first step is to recycle our own urine into the soil, directly or through a compost pile. We're flushing liquid gold into our waterways! – Mike Eaton

"I have a farm that provides food for my family and for my community. I think food is a big part of being prepared, and I think we've got that mostly covered. But I would like to be a better seed saver." – Kathryn Draeger

"I can't pretend that I know what it will look like. I suspect that society will evolve in ways that few of us anticipate. So the greatest skill set will be flexibility and adaptability. The ability to learn and acquire new skills as they become necessary. Things are gonna change in ways that we can't pretend to predict." – Michael Bomford

"Preparing for the future requires increased knowledge and understanding of world events, thinking critically about local and global impacts." – Kelley Eicher

What do you envision as either adaptation or preparation for agriculture and the food system to a restructured society living with less energy?

"We will, sooner or later, recover the skill set necessary to grow and prepare our own food and eat only meats produced incidentally on small, diverse, and nutrient self-sufficient farms and gardens. Foodscapes will be largely local." – Mike Eaton

"Climate change has changed the rainfall patterns where I live in ways we did not predict or expect, it was supposed to get dryer, but instead we're 50% over our historical rainfall averages. I think we're going to have massive variability, and I'm kind of counting on perennial crops—perennial crops, I think, will be part of the solution." – Kathryn Draeger

In places like the U.S. people are very disconnected from food and might have a hard time relating to these topics. Are there ways you have framed the situation that have grabbed people's attention?

"'Know your farmer' is a powerful motivator for those concerned about GMOs, pesticide residues, nutrient density, or climate, and it's a short step from 'know your farmer' to 'be your farmer' (at least in part). The disconnect that we need to fix is manifest at many levels, from the loss of awareness of seasonality to the loss of knowledge about seed saving and about food preparation and preservation. Urban farms and school gardens are powerful change agents because of the large audiences they reach. Slow Food's school garden curricula are important tools, as is their '10,000 school gardens in Africa' campaign as a model for what should be replicated everywhere. 'Farm to Fork' is an important marketing concept but nothing beats 'hand to mouth' for opening a child's eyes to the magical potential of soil, sunlight, and water." – Mike Eaton

"If you can put things in an economic context, like how local food actually preserves, maintains, and even could enhance the wealth of our area, rather than how some people have dubbed the local foods movement "Arugula for All" which is kind of like that elitist, unapproachable system. Instead you can say, 'Look, I'm standing in my grocery store, I have an apple in front of me that's imported from New Zealand, where I have a much better tasting apple right here on the lakeshore that is phenomenally delicious and close, and we support a local family by buying those apples.' That is an argument that resonates." – Kathryn Draeger

"Through discussion with friends and family, I try to help people consider where the food they consume comes from and recognize all the work and process that goes into making food available on grocery store shelves. Consumers often don't realize the total resource and environmental footprint in obtaining raw ingredients, processing, transportation, and end waste of the packaged food they can effortlessly pick up at the grocery store or have delivered to their door. The same is true for any consumed product, not just foods." – Kelley Eicher

"I find that when I go to speak to people, if there's something I can show them—a wheel hoe or a soil probe—that they can look at and play with, people are attracted to that. I find that animals are very charismatic. I can talk about plants, and people are interested, but as soon as I have a goat or a chicken, it brings in 20 times as many people. And that gives an opportunity to talk about the role of livestock in the agro-ecosystems. Those physical teaching tools really attract attention and provide opportunity for discussion." – Michael Bomford

Image Credits

Cover Page image © Peteri/Shutterstock.

Page 3 image courtesy of Christopher Bradford. Angor Wat, Cambodia.

Page 20 image by author. Taken in 2017 near Philomath, Oregon.

Page 28 image courtesy of Christopher Bradford. New York City.

Page 45 image by author. Taken in 2015 near Corvallis, Oregon.

Page 55 image by author. Taken in 2016 near Corvallis, Oregon.

Endnotes

All internet addresses are current as of August-December 2018 unless otherwise indicated.

---

[1] Curtis, Fred, "Peak globalization: Climate change, oil depletion and global trade," (*Ecological Economics* 69, 427-434, 2009).

[2] A good, readable overview of the interrelated systemic issues of climate change in infrastructure, especially water systems, can be found at http://resourceinsights.blogspot.com/2018/09/climate-change-water-and-infrastructure.html.

[3] Cumming, Graeme S. and Garry D. Peterson, "Unifying Research on Social-Ecological Resilience and Collapse," (*Trends in Ecology & Evolution* 32(9): 695-713, 2018); Tainter, Joseph A., *The Collapse of Complex Societies,* (Cambridge University Press, 1988).

[4] Day, John and Charles Hall, *America's Most Sustainable Cities and Regions: Surviving the 21st Century Megatrends,* (Springer, 2016).

[5] Stone, Richard, "Divining Angkor: After rising to sublime heights, the sacred city may have engineered its own downfall," (*National Geographic*, July 2009).

[6] See table and references here: https://en.wikipedia.org/wiki/List_of_largest_cities_throughout_history

[7] Pimentel, David and Marcia H. Pimentel, *Food, Energy, and Society, Third Edition,* (CRC Press, 2007), doi:10.1201/9781420046687.

[8] *Demographia World Urban Areas,* 14th Annual Edition (2018).

[9] Kennedy, Christopher, "A Mathematical Description of Urban Metabolism," pgs 275- 291 in M.P. Weinstein and R.E. Turner (eds.), *Sustainability Science: The Emerging Paradigm and the Urban Environment* (Springer, 2012).

[10] UN DESA Department of Economic and Social Affairs, "World's Population Increasingly Urban with More than Half Living in Urban Areas", (United Nations, 2014).

[11] Folke, Carl, Asa Jansson, Jonas Larsson and Robert Costanza, "Ecosystem Appropriate by Cities," (*Ambio* 26(3): 1997), 167-172.

[12] Many articles can be found espousing this point of view, see for example, Batty, M., "When all the world's a city," (*Environment and Planning A* 43(4), 765-772, 2011). The future projections come from the United Nations Department of Economic and Social Affairs, Population Division. See for example their World Urbanization Prospects 2018 website, factsheets, graphics and reports: https://esa.un.org/unpd/wup/.

[13] United Nations Population Division, World Urbanization Prospects 2018: https://esa.un.org/unpd/wup/.

[14] The cited passage is from page 137 in: Garrett, Timothy J., "Long-run evolution of the global economy: 1. Physical basis," (*Earth's Future* 2: 2014, 217-151). It may at first seem odd that an atmospheric scientist is interested in economic modeling, but it is the growth of the economy via fossil fuel combustion and land use change that is now the primary driver of changes in atmospheric composition.

[15] Heinberg, Richard, and David Fridley. *Our Renewable Future: Laying the Path for 100% Clean Energy.* (Washington: Island Press, 2016), http://ourrenewablefuture.org/.

[16] Center for Sustainable Systems, "U.S. Food System Factsheet," (University of Michigan, 2017), Pub. No. CSS01-06.

[17] Canning, Patrick, Ainsley Charles, Sonya Huang, Karen R. Polenske, and Arnold Waters, "Energy Use in the U.S. Food System," Economic Research Report Number 94 (United States Department of Agriculture, Economic Research Service, 2010).

[18] Center for Sustainable Systems, "U.S. Food System Factsheet," Pub. No. CSS01-06, (University of Michigan, 2017).

[19] For a discussion of energy density and limits to technical advances see: Zenz House, Kurt, "The limits of energy storage technology," (*Bulletin of the Atomic Scientists*, January 20, 2009), https://thebulletin.org/2009/01/the-limits-of-energy-storage-technology. Data for chart from: https://www.engineeringtoolbox.com/fossil-fuels-energy-content-d_1298.html, https://en.wikipedia.org/wiki/Energy_density, and https://en.wikipedia.org/wiki/Energy_content_of_biofuel.

[20] Giaimo, Cara, "When the Western World Ran on Guano," (*Atlas Obscura*, October 14, 2015), https://www.atlasobscura.com/articles/when-the-western-world-ran-on-guano.

[21] Van Kauwenbergh, Steven, *World Phosphate Rock Reserves and Resources*, (Muscle Shoals, AL: IFDC, 2010), https://pdf.usaid.gov/pdf_docs/Pnadw835.pdf.

[22] "Food and Water in an Emergency," (Federal Emergency Management Agency, Washington D.C., August 2004), https://www.fema.gov/pdf/library/f&web.pdf; "Just In Time – JIT," (Investopedia, 2018), https://www.investopedia.com/terms/j/jit.asp.

[23] Plant cultivar adaptation to high fertilizer inputs is an important and until recently overlooked topic that has implications for plant breeding for low input systems, e.g., Murphy, Kevin M., Kimberly G. Campbell, Steven R. Lyon, and Stephen S. Jones, "Evidence of varietal adaptation to organic farming systems," (*Field Crops Research* 102, 172-177, 2007). See also this review paper that connects many topics: Reeve, J.R., L.A. Hoagland, J.J. Villalba, P.M. Carr, A Atucha, C. Cambardella, D.R. Davis, and K. Delate, "Organic Farming, Soil Health, and Food Quality: Considering Possible Links," in *Advances in Agronomy*, Vol 137, 319-368, Academic Press (2016).

[24] Alexander Muller and Pavan Sukhdev, "Measuring What Matters in Agriculture and Food Systems: A synthesis of the results and recommendations of TEEB for Agriculture and Food's Scientific and Economic Foundations Report," (The Economics and of Ecosystems and Biodiversity (TEEB), UN Environment, 2018). Food Tank and Barbara Gemmill-Herren, "Making the Invisible Visible: A Bold Strategy to Change the Food System," (Food Tank, 2018), https://foodtank.com/news/2018/06/teebagrifood-barbara-gemmill-herren/.

[25] Trucost Plc, *Natural Capital at Risk: The Top 100 Externalities of Business,* (TEEB for Business, April 2013). http://naturalcapitalcoalition.org/wp-content/uploads/2016/07/Trucost-Nat-Cap-at-Risk-Final-Report-web.pdf.

[26] University of California – San Diego, "In broadest view yet of world's low oxygen, scientists reveal dangers and solutions," phys.org, January 4, 2018 ,https://phys.org/news/2018-01-broadest-view-world-oxygen-scientists.html.

[27] "Agricultural Practices & Carbon Sequestration Fact Sheet." (Union of Concerned Scientists, 2009), https://www.ucsusa.org/food_and_agriculture/solutions/advance-sustainable-agriculture/ag-carbon-sequest-fact-sheet.html.

[28] Klare, Michael T., *Rising Powers, Shrinking Planet: The New Geopolitics of Energy*, (Metropolitan Books, 2008).

[29] Day, John and Charles Hall, *America's Most Sustainable Cities and Regions: Surviving the 21st Century Megatrends,* (Springer, 2016), page 277.

[30] Day, John and Charles Hall *America's Most Sustainable Cities and Regions: Surviving the 21st Century Megatrends,* (Springer, 2016), pages 310-311.

[31] Internal migration is beginning to be a topic of research, mostly related to climate change and impacts on sea level, major storms, severe drought, and untenable heat. See for example: Goodell, Jeff, "Welcome to the Age of Climate Migration: Extreme weather due to climate change displaced more than a million people from their homes last year. It could soon reshape the nation," (*Rolling Stone*, Feb. 25, 2018), https://www.rollingstone.com/politics/politics-news/welcome-to-the-age-of-climate-migration-202221/. Day and Hall may be unique in accounting for both climate change and energy descent.

[32] The intimate relationship between energy and culture is often called White's Law: https://en.wikipedia.org/wiki/White%27s_law. To explore what future societies may be like, an excellent resource is: Fleming, David and Shaun Chamberlin, *Surviving the Future: Culture, Carnival, and Capital in the Aftermath of the Market Economy,* (Chelsea Green Publishing, 2016).

[33] Holmgren, David, *Feeding Retrosuburbia: from the Backyard to the Bioregion*, (Holmgren Design, 2018).

[34] Masi, Brad, Janet Fiskio and Rumi Shammin, "Urban Agriculture is Rust Belt Cities," (*Solutions* 5(1), 44-53, 2014).

[35] Altieri, Miguel A., Fernando R. Funes-Monzote, "The Paradox of Cuban Agriculture," (*Monthly Review*, January 2012).

[36] Babington, Deepa and Lefteris Papadimas, "In Greece, a painful return to country roots," (*Reuters*, June 7, 2012), https://www.reuters.com/article/us-greece-countryside-idUSBRE85708920120608.

[37] Bomford, Michael, "Getting Fossil Fuels Off the Plate," in Richard Heinberg and Daniel Lerch, eds., *The Post Carbon Reader Managing the 21st Century's Sustainability Crisis,* (Santa Rosa, CA, Post Carbon Institute: 2010), https://www.postcarbon.org/publications/post-carbon-reader/.

[38] Pfeiffer, David Allen, *Eating Fossil Fuels: Oil, Food, and the Coming Crisis in Agriculture*, (New Society Publishers, 2006).

[39] Greer, John Michael, "Progress Vs. Apocalypse: the Stories We Tell Ourselves" in Tom Butler, Daniel Lerch, and George Wuerthner, eds., *The ENERGY Reader: Overdevelopment and the Delusion of Endless Growth*, (Foundation for Deep Ecology, 2012), http://energy-reality.org/progress-vs-apocalypse-the-stories-we-tell-ourselves/.

[40] William Catton termed this "Cargoism" after the quasi-religions that developed in parts of the South Pacific during WWII. More specifically, here is a short video that reviews the current fashions in agriculture and food technologies: "The Future of Farming & Agriculture," (*TheDailyConversation*, YouTube: 2017), https://www.youtube.com/watch?v=Qmla9NLFBvU.

[41] Heinberg, Richard, *There's No App for That* (Corvallis, OR: Post Carbon Institute, 2017), http://www.postcarbon.org/publications/theres-no-app-for-that/.

[42] Pimbert, Michael and Colin Anderson, "The battle for the future of farming: what you need to know," (*The Conversation*, Nov. 21, 2018), https://theconversation.com/the-battle-for-the-future-of-farming-what-you-need-to-know-106805.

[43] "Vertical Farming," (Vertical Farm RSS), http://www.verticalfarm.com/.

[44] "Ethanol Fuel Basics," (U.S. Department of Energy, Alternative Fuels Data Center, undated), https://www.afdc.energy.gov/fuels/ethanol_fuel_basics.html.

[45] "Fertilizer History: The Haber-Bosch Process," (The Fertilizer Institute, 2016), https://www.tfi.org/the-feed/fertilizer-history-haber-bosch-process.

[46] "Farming: There's an App for That," (*National Geographic*, 2018), https://news.nationalgeographic.com/future-of-food/food-future-precision-agriculture/.

[47] See for example, Kowitt, Beth, "Silicon Valley and the Search for Meatless Meat" (*Fortune*, 2017), http://fortune.com/2017/12/19/silicon-valley-meatless-meat/.

[48] European Academies Science Advisory Council, *Negative emission technologies: What role in meeting Paris Agreement targets?* (EASAC policy report 35, February 2018), https://easac.eu/press-releases/details/negative-emission-technologies-will-not-compensate-for-inadequate-climate-change-mitigation-efforts/.

[49] Alcott, Blake, "Jevons' Paradox," (*Ecological Economics* 54: 9-21, 2005).

[50] Ripple, William et al., "World Scientists' Warning To Humanity: A Second Notice," (*BioScience* 67: 2017), 1026-1028.

[51] Wackernagel, Mathis, and William E. Rees, *Our Ecological Footprint: Reducing Human Impact on the Earth*, (Gabriola Island, BC: New Society Publ., 2007); Farley, Joshua, "Ecological Economics," in Richard Heinberg and Daniel Lerch, eds., *The Post Carbon Reader Managing the 21st Century's Sustainability Crisis*, (Santa Rosa, CA, Post Carbon Institute: 2010), https://www.postcarbon.org/publications/post-carbon-reader/.

[52] Meadows, Donnella H., Dennis L. Meadows, Eric Tapley, and Jørgen Randers, *Limits to Growth: The 30-year Update*, (Chelsea Green, 2004). Heinberg, Richard, "The View from Oil's Peak," (2007), http://richardheinberg.com/184-the-view-from-oil's-peak. Lambert, Jessica G., Michelle Arnold, Ajay Gupta, Charles A.S. Hall, and Stephen Balogh, "Energy, EROI and Quality of Life," (*Energy Policy*: 2014), 153-67. Wackernagel, Mathis, and William Rees, *Our Ecological Footprint: Reducing Human Impact on the Earth*, (New Society, 1996). Catton, William R. Jr., *Overshoot: The Ecological Basis of Revolutionary Change*, (Board of Trustees of the University of Illinois, 1980).

[53] Werner, Richard A. "A Lost Century in Economics: Three Theories of Banking and the Conclusive Evidence," (*International Review of Financial Analysis*: 2016), 361-79. doi:10.1016/j.irfa.2015.08.014.

[54] Lerch, Daniel, ed., *The Community Resilience Reader: Essential Resources for an Era of Upheaval*, (Island Press, 2017), https://reader.resilience.org.

[55] Odum, Elisabeth C. and Howard T. Odum, *A Prosperous Way Down: Principles and Policies*, (Boulder: University Press of Colorado, 2008); Hirsch, R. L., Roger Bezdek, and Robert Wendling, "Peaking of World Oil Production: Impacts, Mitigation, & Risk Management," (U.S. Department of Energy, Office of Scientific and Technical Information, 2005), doi:10.2172/939271.

[56] Urgency to act is also considered poor economic policy by many so-called neoclassical economists. This can get highly technical quickly, but in short, they discount the future and believe that by maximizing economic well-being today we will be able to maintain the wealth that allows us to solve the problems of the future through improved technologies and efficiencies. For many not steeped in this narrow field of study, this belief system is considered quite insane. See for example: https://www.resilience.org/stories/2018-11-07/the-secret-of-eternal-growth-its-wishful-thinking/.

[57] Figure from: Walker, Brian and David Salt, "A Crash Course in the Science of Resilience," in Lerch, Daniel, ed, *The Community Resilience Reader: Essential Resources for an Era of Upheaval*, (Island Press, 2017), pgs 163-178.

[58] Walker, Brian and David Salt, "A Crash Course in the Science of Resilience," p 167.

[59] Walker, Brian and David Salt, "A Crash Course in the Science of Resilience," p 173.

[60] For NYC statistics see: https://en.wikipedia.org/wiki/New_York_City.

[61] Images courtesy of Oklahoma Department of Career and Technology Education. https://www.okcareertech.org/educators/cimc/free-samples/ag-cluster/pdf-files/ag2student.pdf/at_download/file.

[62] Pellegrini, Pedro and Roberto J. Fernández, "Crop energy use and the green revolution," (*Proceedings of the National Academy of Sciences* 115(10), 2335-2340, 2018), doi:10.1073/pnas.1717072115.

[63] The term "energy slave" has often been used to help visualize the power of fossil fuels. For a compilation of calculations see: http://energyskeptic.com/2014/energy-slaves/.

[64] Original data from U.S. Energy Information Agency on energy and United Nations DESA/Population Division World Urbanization Prospects on population. Data sets harmonized by author.

[65] Gauthier, Philippe, "The limits of renewable energy and the case for degrowth," (Canadian Dimension, 52(2), 2018), https://canadiandimension.com/articles/view/the-limits-of-renewable-energy-and-the-case-for-degrowth.

[66] "Long-term real growth in US GDP per capita 1871-2009," *VisualizingEconomics*, March 8, 2011, http://visualizingeconomics.com/blog/2011/03/08/long-term-real-growth-in-us-gdp-per-capita-1871-2009.

[67] Perelman, Michael J., "Farming with Petroleum," (*Environment* 14(8), 8-13, 1972).

[68] Fairlie, Simon, *Meat: A Benign Extravagance,* (East Meon, Hampshire: Permanent Publications, 2011).

[69] Montgomery, David R., *Dirt: The Erosion of Civilizations,* (Berkeley, CA: University of California Press, 2012); Diamond, Jared, *Collapse: How Societies Choose to Fail or Succeed,* (Penguin Books, 2011).

[70] A description of the development of the soil texture pyramid can be found here: http://umnsoilsteam.blogspot.com/2014/09/the-historical-development-of-usda.html.

[71] DuPont, S. Tianna, "Soil Quality Information," (Penn State Extension, August 28, 2012), https://extension.psu.edu/soil-quality-information.

[72] Image source: https://commons.wikimedia.org/wiki/File:Soil_profile.jpg.

[73] Jackson, Wes, *Consulting the Genius of the Place: An Ecological Approach to a New,* (Counterpoint Press, 2010); Jackson, Wes, "Tackling the Oldest Environmental Problem: Agriculture and Its Impact on Soil," in Richard Heinberg and Daniel Lerch, eds., *The Post Carbon Reader Managing the 21st Century's Sustainability Crisis,* (Santa Rosa, CA, Post Carbon Institute: 2010), https://www.postcarbon.org/publications/post-carbon-reader/.

[74] Burke, I. C., Yonker, C. M., Parton, W. J., Cole, C. V., Flach, K., & Schimel, D. S., "Texture, Climate, and Cultivation Effects on Soil Organic Matter Content in US Grassland Soils," (*Soil Science Society of America Journal*, 53(3), 1989) 800-805, https://doi.org/10.2136/sssaj1989.03615995005300030029x.

[75] The USDA crop nutrient removal table/calculator can be accessed at: https://plants.usda.gov/npk/main.

[76] Rees, William E., "Cities as Dissipative Structures: Global Change and the Vulnerability of Urban Civilization," pgs 247- 274 in M.P. Weinstein and R.E. Turner, eds., *Sustainability Science: The Emerging Paradigm and the Urban Environment* (Springer, 2012).

[77] King, Franklin H., *Farmers of Forty Centuries: Organic Farming in China, Korea, and Japan*, (Mineola, N.Y: Dover Publications, 2004); Jenkins, Joseph C., *The Humanure Handbook,* (Grove City, PA: Joseph Jenkins Inc., 1996).

[78] As an illustrative example, the daily volume of biosolids from New York City and how it is dealt with is a fascinating topic. See: http://www.nyc.gov/html/dep/html/wastewater/biohome.shtml; http://nymag.com/intelligencer/2018/03/alabamians-are-sick-of-new-yorks-crap.html; https://www.wnycstudios.org/story/poop-train.

79 "Natural Resources Conservation Service," (U.S. Department of Agriculture, NRCS Backyard Conservation, 2018), https://www.nrcs.usda.gov/wps/portal/nrcs/detail/wa/soils/?cid=nrcs144p2_036333.

80 Nearing, M. A., Xie, Y., Liu, B., & Ye, Y., "Natural and anthropogenic rates of soil erosion," (*International Soil and Water Conservation Research*, 2017), https://doi.org/10.1016/j.iswcr.2017.04.001.

81 Diamond, Jared, *Collapse: How Societies Choose to Fail or Succeed*, (Viking Press, 2009).

82 Pretty, Jules and Zareen Pervez Bharucha, "Sustainable intensification in agricultural systems," (*Annals of Botany*, 114(8), 1571–1596, 2014). https://doi.org/10.1093/aob/mcu205.

83 See for example the Oregon State University Central Analytical Laboratory's Soil Health Initiative: https://agsci.oregonstate.edu/central-analytical-laboratory/soil-health-initiative.

84 The USDA educates about soil health through the Natural Resources Conservation Service, see: https://www.nrcs.usda.gov/wps/portal/nrcs/main/soils/health/.

85 Franzluebbers, A. J., Haney, R. L., Hons, F. M., & Zuberer, D. A., "Determination of Microbial Biomass and Nitrogen Mineralization following Rewetting of Dried Soil," (*Soil Science Society of America Journal*, 60(4), 1133-1139, 1996).

86 Montgomery, David R., *Growing a Revolution: Bringing Our Soil Back to Life,* (W.W. Norton & Company, 2017).

87 Pimentel, David and Marcia H. Pimentel, *Food, Energy, and Society*, Third Edition, (CRC Press, 2007).

88 Bender, Martin H, "Energy in Agriculture: Lessons from the Sunshine Farm Project," (Salina, KS: The Land Institute, 2002), http://www.landinstitute.org/vnews/display.v/ART/2001/03/28/3accb0712; Mulder, Kenneth and Ben Dube, "Long-Term Ecological Assessment of Farming Systems (LEAFS): Comparing Human, Animal, and Small Machine Power for Fresh Market Horticulture," *Agroecology and Sustainable Food Systems* 38:6, 704-721 (2014), DOI: 10.1080/21683565.2014.884034.

89 Altieri, Miguel A. and John G. Farrell, *Agroecology: The Science of Sustainable Agriculture,* Second Edition (CRC Press, 2018).

90 See https://commons.wikimedia.org/wiki/File:Tap_and_fibrous_root.jpg.

91 Gold, Mary, "Organic Production/Organic Food: Information Access Tools," (United States Department of Agriculture, National Agricultural Library, June 2007), https://www.nal.usda.gov/afsic/organic-productionorganic-food-information-access-tools.

92 See https://holisticmanagement.org/.

93 See https://www.biodynamics.com/.

94 See https://landinstitute.org/.

95 See http://www.growbiointensive.org/.

96 Holmgren, David, *Permaculture: Principles and Pathways Beyond Sustainability*, (Holmgren Design Services, 2002).

97 For example, see Hathaway, Mark D., "Agroecology and permaculture: addressing key ecological problems by rethinking and redesigning agricultural systems," (*Journal of Environmental Studies and Sciences* 6(2), 239-250, 2016).

98 Walker, Anthony, "Growing Native American Heritage: The Three Sisters," *Poughkeepsie Farm Project*, 2016, https://www.farmproject.org/blog/2016/3/31/growing-native-american-heritage-the-three-sisters.

[99] Walker, Anthony, "Growing Native American Heritage: The Three Sisters". The figure is by Mardi Dodson, from a concept taken from Michael Caduto and Joseph Cruchac, *Native American Gardening: Stories, Projects, and Recipes for Families*, (Fulcrum, 1996).

[100] Public policy recommendations and potential system "lock in" issues are covered well in: Neff, Roni A, Cindy L. Parker, Frederick L. Kirschenmann, Jennifer Tinch, and Robert S. Lawrence, "Peak Oil, Food Systems, and Public Health," (*American Journal of Public Health* 101(9), 1587-1597, 2011).

[101] For a heartbreaking example of what can go wrong and how governments can resist needed change look at the crisis in Venezuela, e.g., Kohut, Meredith and Isayen Herrera, "As Venezuela Collapses, Children Are Dying of Hunger," (*New York Times*, 2017), https://www.nytimes.com/interactive/2017/12/17/world/americas/venezuela-children-starving.html.

[102] Dvoskin, Dan and Earl O. Heady, *U.S. agricultural production under limited energy supplies, high energy prices, and expanding agricultural exports,* (CARD Reports 64, Center for Agriculture and Rural Development, Iowa State University, 1976).

[103] "World Fertilizer Prices Drop Dramatically After Soaring to All-time Highs," (EurekAlert!, 2008), https://www.eurekalert.org/pub_releases/2008-12/i-wfp121608.php. "Managing Cover Crops Profitably, 3rd Edition," (Sustainable Agriculture Research and Education, 2007), . https://www.sare.org/Learning-Center/Books/Managing-Cover-Crops-Profitably-3rd-Edition/Text-Version.

[104] See the "Shale Reality Check" series of reports by earth scientist David Hughes here: http://shalebubble.org/.

[105] Searching for news reports on trucker strikes reveals several examples of temporary shortages in stores due to problems with distribution, e.g., https://www.theguardian.com/business/2008/jun/12/oil.spain; http://news.bbc.co.uk/2/hi/europe/7445758.stm.

[106] Famiglietti, J. S., M. Lo, S. L. Ho, J. Bethune, K. J. Anderson, T. H. Syed, S. C. Swenson, C. R. de Linage M. Rodell, "Satellites measure recent rates of groundwater depletion in California's Central Valley," (*Geophysical Research Letters*, 38(3), 2011).

[107] Perez, Ines, "Climate Change and Rising Food Prices Heightened Arab Spring," (*Scientific American*, 2018), https://www.scientificamerican.com/article/climate-change-and-rising-food-prices-heightened-arab-spring/.

[108] A much more thorough treatment of many topics and examples given in this report can be found in the excellent book, Ackerman-Leist, Phillip, *Rebuilding the Foodshed: How to Create Local, Sustainable, and Secure Food Systems*, (Chelsea Green Publishing, 2013). Also, this review of food system assessment tools may provide helpful guidance: Freedgood, J., Pierce-Quiñonez, M., & Meter, K. A., "Emerging assessment tools to inform food system planning," (*Journal of Agriculture, Food Systems, and Community Development*, 2(1), 83–104, 2011). http://dx.doi.org/10.5304/jafscd.2011.021.023.

[109] Walker, Brian and David Salt, "A Crash Course in the Science of Resilience," p 176.

[110] Parker, Lauren E and John T Abatzoglou, "Projected changes in cold hardiness zones and suitable overwinter ranges of perennial crops over the United States," (Environmental Research Letters 11: 1-8, 2016).

[111] Oborne, A, Mertens, M, "Oregon Food Infrastructure Gap Analysis," (Portland, OR: Ecotrust, 2015).

[112] Peters, Christian J., Jamie Picardy, Amelia F. Darrouzet-Nardi, Jennifer L. Wilkins, Timothy S. Griffin, and Gary W. Fick, "Carrying Capacity of U.S. Agricultural Land: Ten Diet Scenarios," (*Elementa: Science of the Anthropocene* 4, 2016), 000116, doi:10.12952/journal.elementa.000116.

113 Davis, A. S., Hill, J. D., Chase, C. A., Johanns, A. M., & Liebman, M., "Increasing Cropping System Diversity Balances Productivity, Profitability and Environmental Health," PLoS ONE 7(10): e47149 (2012). https://doi.org/10.1371/journal.pone.0047149.

114 See https://www.retrosuburbia.com/reading/feeding-retrosuburbia/.

115 Mulder, Kenneth and Ben Dube, "Long-Term Ecological Assessment of Farming Systems (LEAFS): Comparing Human, Animal, and Small Machine Power for Fresh Market Horticulture," *Agroecology and Sustainable Food Systems* 38:6, 704-721 (2014), DOI: 10.1080/21683565.2014.884034.

116 Heinberg, Richard and Michael Bomford, *The Food and Farming Transition: Toward a Post Carbon Food System*, (Sebastopol, CA: Post Carbon Institute, 2009).

117 Heinberg and Bomford, *The Food and Farming Transition.*

118 An example of a business that encourages fermented food preservation techniques, which have low embodied energy needs: https://www.culturesforhealth.com/.

119 See for example this project: https://nofoodleftbehindcorvallis.org/.

120 Harper, Alethea, Annie Shattuck, Eric Holt-Giménez, Alison Alkon and Frances Lambrick, *Food Policy Councils: Lesson Learned,* (Food First, Institute for Food and Development Policy, 2009).

121 See https://www.nrcs.usda.gov/wps/portal/nrcs/site/national/home/.

122 To find the extension office near you, see this website of the agency (U.S. Department of Agriculture, National Institute of Food and Agriculture) that funds much of the extension work in the United States: https://nifa.usda.gov/extension.

123 See https://www.sare.org/.

124 See http://www.nacdnet.org/.

125 See the work of the Slow Money network: https://slowmoney.org/.

126 See for example: Agriculture & Land-Based Training Association, http://www.albafarmers.org/; Agriculture of the Middle, http://agofthemiddle.org/ and https://ecotrust.org/project/ag-of-the-middle/; Intervale Center, https://www.intervale.org/; School of Adaptive Agriculture, https://www.school-of-adaptive-agriculture.org/.

127 See for example the work on getting local produce into rural grocery stores in Minnesota: https://www.minnpost.com/economic-vitality-in-greater-minnesota/2018/09/beyond-farmers-markets-helping-small-growers-get-fresh-produce-into-grocery-stores/.

128 See resources such as: http://sustainablefarmlease.org/; http://www.savannainstitute.org/land-access.html.

129 Many resources exist to support school garden programs, for example: https://edibleschoolyard.org/; http://www.lifelab.org/; https://www.slowfoodusa.org/school-gardens; https://kidsgardening.org/.

130 See http://www.farmtoschool.org/.

131 See https://foodcorps.org/.

132 See https://www.fns.usda.gov/ebt/snap-and-farmers-markets.

133 See https://communitygarden.org/.

134 Bryce, Emma, "Metropolitan areas are becoming significant sources of sustainable produce," *Anthropocene*, October 2018. http://www.anthropocenemagazine.org/2018/10/metropolitan-areas-are-becoming-significant-sources-of-sustainable-produce/.

[135] Prentice, Jessica, *Full Moon Feast: Food and the Hunger for Connection*, (White River Junction, VT: Chelsea Green, 2006).

[136] See https://www.slowfood.com/; https://www.slowfoodusa.org/.

[137] See for example: https://civileats.com/2016/08/29/forget-farm-to-table-its-farm-to-institution-sourcing-that-could-make-a-real-dent-the-food-system/; https://www.nal.usda.gov/afsic/farm-institution; https://www.farmtoinstitution.org/; http://www.bamco.com/timeline/farm-to-fork/.

[138] To learn from people in the U.S. who are working through the practicalities of living without fossil fuels see: http://www.livingenergyfarm.org/.

[139] Freedgood, J., Pierce-Quiñonez, M., & Meter, K. A., "Emerging assessment tools to inform food system planning," (*Journal of Agriculture, Food Systems, and Community Development* 2(1), 83–104, 2011). http://dx.doi.org/10.5304/jafscd.2011.021.023.

[140] See supplementary materials for this report at https://www.postcarbon.org/future-is-rural.

[141] Fairlie, Simon, "Can Britain Feed Itself?," (The Land 4: 18-26, 2007).

[142] Peters, Christian J., Jamie Picardy, Amelia F. Darrouzet-Nardi, Jennifer L. Wilkins, Timothy S. Griffin, and Gary W. Fick, "Carrying Capacity of U.S. Agricultural Land: Ten Diet Scenarios," (*Elementa: Science of the Anthropocene* 4, 2016), 000116, doi:10.12952/journal.elementa.000116.

[143] A good example of the kind of detailed spatial analyses possible is: Galzki, J. C., Mulla, D. J., & Meier, E., "Mapping potential foodsheds using regionalized consumer expenditure data for Southeastern Minnesota," (*Journal of Agriculture, Food Systems, and Community Development* 7(3), 181–196, 2017). http://dx.doi.org/10.5304/jafscd.2017.073.013.

[144] Delate, Kathleen & Cambardella, Cynthia & Chase, Craig & Turnbull, Robert, "A Review of Long-Term Organic Comparison Trials in the U.S.," (*Sustainable Agriculture Research* 4(3), 5-14, 2015). doi:10.5539/sar.v4n3p5.

[145] See for example the IPCC special report released in October 2018: http://www.ipcc.ch/report/sr15/.

[146] Tallaksen, J., Bauer, F., Hulteberg, C., Reese, M., & Ahlgren, S., "Nitrogen fertilizers manufactured using wind power : greenhouse gas and energy balance of community-scale ammonia production," (*Journal of Cleaner Production* 107: 626-635, 2015). doi:10.1016/j.jclepro.2015.05.130.

[147] Bardi, U., El Asmar, T., & Lavacchi, A., "Turning electricity into food: the role of renewable energy in the future of agriculture," (*Journal of Cleaner Production* 53, 224–231, 2015). doi:10.1016/j.jclepro.2013.04.014.

[148] A comprehensive description of these methods from a diversity of places is the subject of the book: Montgomery, David R., *Growing a Revolution: Bringing Our Soil Back to Life,* (W.W. Norton & Company, 2017).

[149] A gateway into the spatial data available for farm soils in the U.S. is the Web Soil Survey: https://websoilsurvey.sc.egov.usda.gov/App/HomePage.htm.

[150] Rockoff, Hugh, *The U.S. Economy in WWII as a Model for Coping with Climate Change,* (National Bureau of Economic Research, Working Paper 22590, September 2016). https://www.nber.org/papers/w22590.pdf.

[151] A good example of how easy it can be to burst the hype and then point to the underlying issues being ignored is: Fridley, David and Richard Heinberg, "Can Climate Change be Stopped by Turning Air Into Gasoline?" *Renewable Energy World*, June 2018, https://www.renewableenergyworld.com/ugc/articles/2018/06/14/can-climate-change-be-stopped--by-turning-air-into-gasoline.html.

[152] USDA Economic Research Service, using data from USDA, National Agricultural Statistics Service, Censuses of Agriculture (through 2012) and *Farms and Land in Farms: 2016 Summary*.

[153] For a discussion of the inadequacy of individualism see: https://www.theguardian.com/environment/true-north/2017/jul/17/neoliberalism-has-conned-us-into-fighting-climate-change-as-individuals.

Made in United States
Orlando, FL
04 November 2026

90208511R00067